Ultrafast All-Optical Signal Processing Devices

Ultrafast All-Optical Signal Processing Devices

Edited by

Hiroshi Ishikawa

National Institute of Advanced Industrial Science and Technology (AIST), Japan

WILEY

A John Wiley and Sons, Ltd, Publication

This edition first published 2008
© 2008 John Wiley & Sons Ltd

Registered office
John Wiley & Sons Ltd, The Atrium, Southern Gate, Chichester, West Sussex, PO19 8SQ,
United Kingdom

For details of our global editorial offices, for customer services and for information about how
to apply for permission to reuse the copyright material in this book please see our website at
www.wiley.com.

Library of Congress Cataloging in Publication Data

Ultrafast all-optical signal processing devices / edited by Hiroshi Ishikawa.
     p.   cm.
  Includes bibliographical references and index.
  ISBN 978-0-470-51820-5 (cloth)
  1. Optoelectronic devices.   2. Very high speed integrated circuits.
  3. Signal processing—Equipment and supplies.   4. Integrated optics.
  5. Optical data processing.   I. Ishikawa, Hiroshi.
  TK8304.U46 2008
  621.382′2—dc22
                                          2008013161

A catalogue record for this book is available from the British Library.

ISBN 978-0-470-51820-5 (HB)

Typeset in 10/12pt Times by Integra Software Services Pvt. Ltd, Pondicherry, India
Printed in Singapore by Markono Print Media Pte Ltd

Contents

Contributors

Ryoichi Akimoto,
Ultrafast Photonic Devices Laboratory, National Institute of Advanced Industrial Science and Technology (AIST), Ibaraki, Japan.

Hiroshi Ishikawa (editor)
Ultrafast Photonic Devices Laboratory, National Institute of Advanced Industrial Science and Technology (AIST), Ibaraki, Japan.

Hiroshi Ito,
Center for Natural Sciences, Kitasato University, Kanagawa, Japan

Satoshi Kodama,
NTT Photonics Laboratories, NTT Corporation, Kanagawa, Japan

Haruhiko Kuwatsuka,
Nanotechnology Research Center, Fujitsu Laboratories Ltd, Atsugi, Japan.

Hitoshi Murai,
Networks and Devices Laboratories, Corporate R&D Center, Oki-Electric Company, Tokyo, Japan

Shigeru Nakamura,
Nano-electronics Research Laboratory, NEC Corporation, Ibaraki, Japan

Yoh Ogawa,
Networks and Devices Laboratories, Corporate R&D Center, Oki-Electric Company, Tokyo, Japan

Nobuo Suzuki,
Corporate Research and Development Center, Toshiba Corporation, Kawasaki, Japan

Hidemi Tsuchida,
Photonics Research Institute, National Institute of Advanced Industrial Science and Technology (AIST), Ibaraki, Japan.

Preface

This book describes the up-to-date research and development of semiconductor-based, ultra-fast, all-optical, signal processing devices for transmission systems in the range of 100 Gb/s to 1 Tb/s. The contents of the book are based on the Tutorial Presentation in ECOC (European Conference on Optical Communications) 2006 at Cannes, France, entitled *Ultrafast Devices for OTDM Systems* by the present editor. Many researchers in Japan provided their precious materials for the presentation and the editor asked these researchers to be the contributors to this book.

Owing to the recent spread of broadband networks, we can enjoy various services from the network. However, recent rapid increases in communication traffic are causing a serious problem, namely the large power consumption of the network equipment. One possible solution to this problem is to realize ultrafast systems that can transmit a huge amount of data with minimum wavelength division multiplexing (WDM) and minimum conversion of optical signals to electric signals. This holds provided low-power-consuming ultrafast signal processing all-optical devices are realized. The motivation for research and development of ultrafast, all-optical, signal processing devices is to construct such low-power consuming ultrafast networks, thus providing high capacity and real-time information communications. Benefits will be, for example, high reality TV conferences, remote presence, entertainments, remote diagnosis and medical treatment based on high resolution real time pictures, and access to the abundant data and computer resources distributed all over the world. Such networks are also indispensable for our economy and production.

So far, extensive research and development have been done to realize all-optical signal processing devices for a bit-rate of 100 Gb/s to 1 Tb/s, both on fiber-based devices and semiconductor-based devices. In this book, however, we focus on the semiconductor-based devices because of their small size and the feasibility of integration with other semiconductor devices for higher functionality. We believe that realization of semiconductor-based devices is a prerequisite for the commercial ultrafast network systems, where criteria are the cost and the size of equipment once required performance is satisfied.

In ultrafast, all-optical devices, optical nonlinearity is used as the operating mechanism. There is an intrinsic trade-off relationship in that a faster all-optical device requires greater optical power for operation. Efforts have been made under this restriction to develop low power consumption ultrafast devices. In this book, we describe light sources, various types of all-optical gate devices, and wavelength converters, where new ideas and concepts are challenged. We also review recent ultrafast transmission experiments to see the trend of system researches, and to consider the further issues to be overcome in such devices to make the ultrafast systems into real ones. The reader will be able to see the up-to-date challenges in developing semiconductor-based, ultrafast all-optical, signal processing devices in this book.

As this book is based on the contributions of ten researchers in this field, there are some differences in notation and terminology depending on the contributor. The editor apologizes for this; however, the book is so edited that each chapter can be read independently so this should not cause inconvenience to the readers.

Finally on behalf of the contributors to this book, the editor acknowledges the many researchers who worked together with the contributors. Also acknowledged are the funding agencies for the various projects that contributed largely to the progress of the devices described in this book.

Hiroshi Ishikawa

1

Introduction

Hiroshi Ishikawa

1.1 Evolution of Optical Communication Systems and Device Technologies

Deployment of the optical communication systems started at the end of 1970s. The bit rate of the early-stage systems was 100 Mb/s (1980), increasing to 400 Mb/s, 565 Mb/s, 1.6 Gb/s, 2.4 Gb/s and 10 Gb/s over the past three decades. Increasing the transmission capacity was achieved not just by increasing the bit rates as wavelength division multiplexing (WDM) technology was developed in the 1990s. Systems capable of 100–200 wavelengths multiplexing with a single channel bit rate of 2.4 Gb/s and 10 Gb/s were deployed, having scalable total capacities up to 2 Tb/s. Recently, deployments of WDM systems with a single channel bit rate of 40 Gb/s have started. Looking into the future, we will be required to realize still larger capacity networks, as will be discussed later.

Owing to the above-mentioned increase in transmission capacity, broadband Internet network systems have come to be used widely since we entered Twenty-first century. Internet protocols (IP), various browsing technologies, varieties of related software, and increased performance of personal computers and routers, largely contributed to the spread of broadband networks, which have had a huge impact on our society and our daily life. Worldwide e-commerce and e-business has become an essential part of our economy with outsourcing of office jobs, research and development being done using networks. Even production at remote sites is becoming possible though networks. The world-wide impact of broadband networks is clearly described in such books as *Revolutionary Wealth* by Alvin Toffler and Heidi Toffler[1], and *The World is Flat* by Thomas L. Friedman[2].

When we looked back the technological evolution of these networks, development of new or higher performance devices and components played crucial roles. Such devices were low-loss optical fibers, semiconductor lasers, detectors such as APDs (avalanche photodiodes) and PIN photodiodes, integrated driver circuits, multiplexing and demultiplexing ICs, and fiber

---

Ultrafast All-Optical Signal Processing Devices   Edited by Hiroshi Ishikawa
© 2008 John Wiley & Sons, Ltd

amplifiers. Many passive components such as arrayed waveguide gratings (AWG) and optical filters were needed for WDM systems.

We can see a good example in light sources showing how their innovation contributed to an increase in transmission bit rates. First, Fabry-Perot lasers, which lased in multiple spectra enabled transmission rates of up to 400 Mb/s. To increase the bit rate to more than 1 Gb/s, lasers with single wavelengths were essential to minimize the effect of chromatic dispersion of fiber. Distributed feedback (DFB) lasers were developed to this end. For longer-span transmission with bit rates above 10 Gb/s, wavelength chirp in a single lasing spectrum was a problem. Then the external modulation scheme was developed. Electro-optic modulators using LiNbO$_3$, semiconductor-based, electro-absorption modulators (EAM) were developed. Monolithic integration of DFB laser and EAM was done to realize a compact light source. Owing to these advances in light sources together with advances in other devices and technologies, it was possible to increase the transmission bit rate. If we target much higher bit rate systems, such as 100 Gb/s and 1 Tb/s, for future applications, the key will be the development of new and higher performance devices as well.

1.2 Increasing Communication Traffic and Power Consumption

Figure 1.1 shows the long term trend of communication traffic in JPIX, which is one of the major Internet exchangers in Tokyo. The traffic in JPIX is increasing by 40 to 50 % per year. Figure 1.2 shows the total traffic in Japan. The plots using closed circles are the time-averaged amount of information per second being downloaded from networks as announced by MIC (Ministry of Internal Affairs and Communications). The value was 324 Gb/s at November 2004. This increased to 722 Gbps in May 2007. The solid line in the figure is the estimated total traffic assuming a 40 % annual increase. One of the driving forces for this rapid increase is the increase in subscribers to broadband. Initially, ADSL (Asymmetrical Digital Subscriber Line) was used; however, recently the FTTH (Fiber to the Home) subscribers are increasing

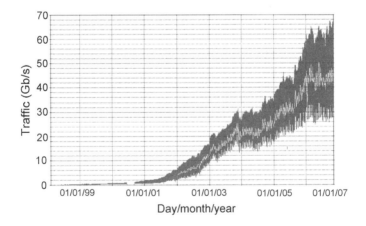

Figure 1.1   Internet traffic in JPIX, which is one of the major Internet exchanges in Tokyo. (Reproduced by permission of JPIX. (http://www/jpix.ad.jp/techncal/traffic.html))

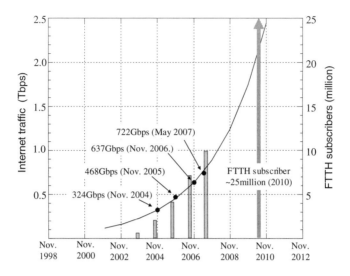

Figure 1.2 Total traffic in Japan. Solid circles are evaluated value by Ministry of Internal Affairs and Communications. The line is the fit assuming a 40 % annual increase. Bars show the subscribers to FTTH

rapidly. NTT, one of the major carrier companies in Japan, is aiming at 20 million subscribers to FTTH by 2010. The bars in Fig. 1.2 are subscribers to FTTH. NTT is to bring NGN (Next Generation Network) into service in 2008. NGN is an IP-based network enabling various services with higher quality [3].

The dramatic increase in traffic and the plan to increase various services will cause a serious problem, namely the power consumption of the network equipment. Figure 1.3 shows the router power consumption in Japan as estimated by T. Hasama of AIST (National Institute of Advanced Industrial Science and Technology). The power consumption in 2001 is based on actual data. Assuming a 40 % annual increase in traffic and reduction of the CMOS-LSI drive voltage, plotted as closed circles in the figure, the power consumption of routers will reach 6.4 % of the total power generation in 2020 even for the low CMOS-LSI drive voltage of 0.8 V. If the drive voltage reduction of CMOS-LSI is insufficient, the power consumption will still easily reach a few tens of a percentage point or more. This means we cannot have the benefits of larger capacity networks.

One of the causes of large power consumption in the present network is the WDM scheme and electrical routing of the packet signals. The WDM requires O/E (optical to electrical) signal and E/O (electrical to optical) signal conversion circuits with the same number as that of the wavelength, resulting in an increase in power consumption. In addition to this, electrical signal processing for IP packet routing and switching at the router consumes large amounts of power. If we could realize 100 Gb/s to 1 Tb/s bit-rate transmission, huge capacity data could be transmitted with a small number of wavelengths, which might reduce the power consumption. If we could process ultrafast signals without converting to electrical signals, this would also reduce the power consumption of routers. Consequently, the development of ultrafast all-optical devices is very important for future, low power-consumption huge capacity networks

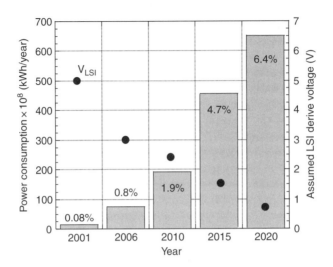

Figure 1.3 Estimated power consumption by Internet routers in Japan. Original data is from T. Hasama of AIST. The value for 2001 is actual data. Plots shown by solid circles are the assumed drive of LSI voltage used in routers. The percentages in the figure are the proportion of total power generation. If we assume a 40 % increase in traffic, the router power consumption reaches 6.4 % of the total power generation in 2020 even for a low LSI drive voltage of 0.8 V

1.3 Future Networks and Technologies

1.3.1 Future Networks

Forecasting the future of networking is of large importance in planning research and development. It is obvious that the traffic of video content will keep on increasing. At present, a large proportion of the network bandwidth is occupied by video content, such as TV and movies, and moving-picture distribution services. The convergence of broadcasting and communication will soon take place in NGN (Next Generation Network). Network users will require higher resolution pictures; however there is a limitation on resolution due to the limited bandwidth. International distribution of 4K-digital cinema (for resolution of 2000×4000, the required bandwidth is above 6 Gb/s without compression) by network was demonstrated using data compression by JPEG2000 [4]. NHK (the Japanese public broadcasting organization) is developing ultra-HDTV (high definition TV) having a resolution of 4320×7680, requiring a bandwidth of 72 Gb/s, and is planing to start broadcasting ultra-HDTV in 2025 [5].

If we could get rid of the bandwidth limitation, there would arise a lot of new applications. Higher resolution, real-time, moving pictures with realistic sound will make the TV conference a far more useful tool. International conferences could even be held using remote-presence technology. This would reduce the energy consumption by reducing the traffic. Medical applications for the network will also be important. Using high-resolution pictures without time delay, remote diagnosis can be done, and even remote surgery is within its scope. Other important associated technology would be grid technology. One of the present applications of grid technology is to establish connections or paths between various computer sites or data storages. The large bandwidth optical paths, which can be controlled by a user, will enable high performance grid computing (e-science) by connecting computers worldwide. Grid-based virtual huge capacity storage, and grid-based economy (e-economy, e-commerce, e-production) will also be important issues.

When we look at the current IP-based network, it is not suited to handling such a huge capacity of data. It is optimized rather to low granularity traffic and requires data compression for large-capacity data because of the bandwidth limitation. This causes the time delay, and we cannot obtain the benefit of real-time information. A novel network capable of real-time, high-capacity, transmission is required. One candidate is the optical-path network, in which end-to-end connection and broadcasting end to multi-ends connection can be achieved with optical paths where large-scale optical switches are used for routing. The concept of optical path has been discussed in terms of wavelength path or virtual wavelength path [6]. The dynamic huge-scale path network including the wavelength path with very high bit rate is highly attractive as a future network for huge capacity data transmission. In such path systems, information can be transmitted transparently, i.e. regardless of the modulation format and bit rate, without using electronic routers. Combination of IP based networks, which handle small granularity data, and dynamic optical path networks, which handle huge capacity data with very high bit rates, is one of the promising forms of the future network.

1.3.2 Schemes for Huge Capacity Transmission

There are two ways of achieving huge capacity transmission. One is the optical time division multiplexing (OTDM) technology as illustrated in Figure 1.4, and the other is the employment of multilevel modulation schemes as illustrated by Figure 1.5.

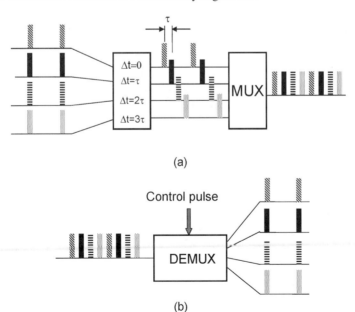

(a)

(b)

Figure 1.4 Optical time division multiplexing (OTDM) scheme. (a) By giving proper delay to each channel we can generate a very fast optical signal. (b) For demultiplexing we are required to develop all-optical switches

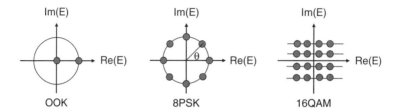

Figure 1.5 Constellation diagram of OOK (on–off keying), 8 PSK (phase shift keying) and 16 (quadrature amplitude modulation). By utilizing phases of light waves we can realize multilevel modulation

In the OTDM scheme, optical signals from different channels are multiplexed by applying a proper delay to each channel in order to get high bit-rate signals. We can generate high bit rates, for example 160 Gb/s or 1.28 Tb/s [7], which cannot be achieved by electric circuits. To make the OTDM systems into real ones, we need to develop ultrafast, all-optical signal processing devices. There are ultrafast light sources and ultrafast all-optical gate switches for such functions as gating, clock extraction, 2R (retiming and reshaping) operations, and DEMUX (demultiplexing). To make the system flexible, a wavelength conversion device is also essential. Dispersion compensation including polarization-mode dispersion, is also an important issue for long-distance transmission.

The other scheme involves the use of multilevel modulation, which not only uses the amplitude of light but also the phase [8, 9]. By utilizing phases of the light field we can perform multilevel modulation. Figure 1.5 shows examples of multilevel modulation in the form of constellation mapping. The horizontal axis is the real part of the electric field and the vertical axis is the imaginary part. Figure 1.5(a) is the conventional on–off keying (OOK). Figure 1.5(b) is 8 PSK (phase shift keying), which can transmit 3 bit/symbol, and (c) is 16 QAM (quadrature amplitude modulation) capable of 4 bit/symbol modulation. Precise control of phases and sophisticated decoding technology are required to realize a large multilevel [10, 11]. The multilevel scheme has an advantage in that it can increase the total capacity without increasing the symbol rate. This makes the dispersion compensation easier.

1.4 Ultrafast All-Optical Signal Processing Devices

1.4.1 Challenges

In this book we describe the challenges for semiconductor-based ultrafast (100 Gb/s - 1 Tb/s) all-optical signal processing devices. A major application is in ultrafast OTDM networks; however, a multi-level scheme based on a symbol rate beyond 100 Gb/s could also be a possibility in further increasing the transmission rate. Focus is put on semiconductor-based devices, although fiber-based devices are used for ultrafast OTDM experiments, for example, NOLM (Nonlinear Optical Loop Mirror) [12, 13]. Advantages of semiconductor devices when compared with fiber devices are their small size and possible integration of devices for higher functionality. With semiconductor devices, however, there is a lot of difficulties in realizing practical devices. One of the major difficulties is the intrinsic one that faster all-optical device operation based on optical nonlinearity requires larger optical energy. This is theoretically illustrated in the next section. This problem can be avoided in fiber devices because long fiber-lengths can be used to obtain sufficient nonlinearity for low energy operation. In semiconductor devices, although the nonlinear susceptibility is greater than with optical fibers, device sizes are very small. It is not, therefore, easy to realize low-energy operating devices; hence, for the development of ultrafast all-optical semiconductor devices, full utilization of many new ideas and concepts are required.

A systematic challenge for semiconductor-based, ultrafast all-optical devices was The Femtosecond Technology Project (1995–2004) in Japan, which was conducted with the support of the Ministry of Trade and Industry, and NEDO (New Energy and Industrial Technology Development Organization) [14]. Mode-locked semiconductor lasers were developed, as were various types of all-optical gate switches, and WDM transmission technology based on 160 Gb/s–320 Gb/s OTDM signals. Described in this book are mode-locked lasers (Chapter 2), symmetric Mach–Zehnder gate switch (Chapter 3), intersub-band transition gate switches (Chapter 5), four-wave mixing wavelength converters (Chapter 6), and transmission technologies (Chapter 7). Another project, named 'Research and Development on Ultrahigh-speed Backbone Photonic Network Technologies' (1996–2005) was conducted under the auspices of NICT (National Institute of Information and Communication Technology). In this project, a 160-Gb/s CS-RZ (carrier suppressed return to zero) signal was generated by OTDM technology using an electro-absorption modulator (EAM) [15]. A field transmission experiment was demonstrated over 635 km. The OTDM light source developed in this project is described in Chapter 2, and the transmission experiment is briefly reviewed in Chapter 7. Outside of these projects, much interesting research work has been done worldwide, including a device using an ultrafast photodiode and traveling-wave electro-absorption modulator, described in Chapter 4, and a use of SOA with wavelength filter enabling use of only the very fast response component of SOA response (Chapter 3).

1.4.2 Basics of the Nonlinear Optical Process

For ultrafast, all-optical, signal processing using semiconductor-based devices, we use optical nonlinear effects, mainly the third-order nonlinearity. The third-order process is highly useful since it gives such effects as absorption saturation (gain saturation) and four-wave mixing. Here we briefly look at the third-order nonlinear process, taking the simplest two-level system as an example in order to achieve basic understanding of the device operation and to illustrate the intrinsic difficulty with all-optical ultrafast devices.

Figure 1.6 shows a two-level system. We assume N two-level systems with inversion symmetry in a volume V. We consider a case where only one frequency plane wave with angular frequency ω is applied. The response of the two-level system to the optical field can be described by a density matrix equation of motion [18, 19]. If we write down all the components of the equation of motion:

$$\frac{d}{dt}\rho_{aa} = \frac{i}{\hbar}(\rho_{ab}H_{ba} - H_{ab}\rho_{ba}) - \gamma_a\left(\rho_{aa} - \rho_{aa}^{(0)}\right) \tag{1.1}$$

$$\frac{d}{dt}\rho_{bb} = \frac{i}{\hbar}(\rho_{ba}H_{ab} - H_{ba}\rho_{ab}) - \gamma_b\left(\rho_{bb} - \rho_{bb}^{(0)}\right) \tag{1.2}$$

$$\frac{d}{dt}\rho_{ab} = \frac{i}{\hbar}(E_b - E_a)\rho_{ab} + \frac{i}{\hbar}(H_{ab}\rho_{aa} - H_{ab}\rho_{bb}) - \gamma_{ab}\rho_{ab} \tag{1.3}$$

$$\frac{d}{dt}\rho_{ba} = \frac{i}{\hbar}(E_a - E_b)\rho_{ba} + \frac{i}{\hbar}(H_{ba}\rho_{bb} - H_{ba}\rho_{aa}) - \gamma_{ab}\rho_{ba} \tag{1.4}$$

For a plane-wave electric field $\mathbf{E}(\omega)$, the perturbation Hamiltonian under dipole approximation can be written as,

$$H_{ab}(\omega) = -\boldsymbol{\mu}_{ab} \cdot \mathbf{E}(\omega) = -\boldsymbol{\mu}_{ab} \cdot (\mathbf{E}_{\omega}e^{-i\omega t} + c.c.) \tag{1.5}$$

where $\boldsymbol{\mu}_{ab}$ is a dipole moment given by:

$$\boldsymbol{\mu}_{ab} = \langle a|e\mathbf{r}|b\rangle \tag{1.6}$$

$\rho_{aa}^{(0)}$ and $\rho_{bb}^{(0)}$ are the unperturbed diagonal elements of the density matrix, which can be replaced by electron distribution functions such as Fermi–Dirac or Boltzmann distribution function under thermal equilibrium. γ_a and γ_b are the phenomenological relaxation rates of the diagonal component of the thermal equilibrium. We may put $\gamma = \gamma_a = \gamma_b = 1/T_1$, where T_1 is the energy relaxation time of an electron. γ_{ab} is the dephasing rate of the off-diagonal element. Elastic scattering, as well as inelastic scattering, of electrons contributes to the dephasing of a dipole. Its inverse is the dephasing time T_2.

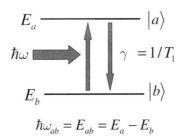

Figure 1.6 A model of two-level system

The equation of motion can be solved by using the iterative procedure. A first-order solution for the off-diagonal component can be obtained by using unperturbed diagonal terms for the right-hand side of Equation (1.3). Retaining only the resonant term, we obtain:

$$\rho_{ab}^{(1)}(\omega) = -\frac{\boldsymbol{\mu}_{ab} \cdot \mathbf{E}_{\omega}e^{-i\omega t}}{\hbar(\omega_{ab} - \omega) - i\hbar\gamma_{ab}}(\rho_{aa}^{(0)} - \rho_{bb}^{(0)}) + cc. \tag{1.7}$$

Inserting this into Equations (1.1) and , we obtain second-order solution, $\rho_{aa}^{(2)}$ and $\rho_{bb}^{(2)}$. Again, by inserting second-order diagonal terms into Equation (1.3), we obtain the third-order solution of the off-diagonal term. This is a rather tedious calculation procedure. Retaining only the resonant terms makes the analysis simpler. Once the density matrix components are known, the polarization of the system is given by:

$$\mathbf{P}(\omega) = \frac{2N}{V}Tr(\boldsymbol{\mu}\rho) = \frac{2N}{V}(\boldsymbol{\mu}_{ab}\rho_{ab} + \boldsymbol{\mu}_{ba}\rho_{ba}) \tag{1.8}$$

The first-order solution of the off-diagonal terms gives the first-order polarization, and the third-order off-diagonal terms give the third-order polarization. Retaining the first- to the third-order terms, the generated polarization can be written as:

$$\mathbf{P}(\omega) = \varepsilon_0 \left(\chi^{(1)}(\omega) + \chi^{(3)}(\omega) |\mathbf{E}_\omega|^2 \right) \mathbf{E}_\omega \tag{1.9}$$

where ε_0 is the vacuum dielectric constant, $\chi^{(1)}$ is linear susceptibility and $\chi^{(3)}$ is the third-order nonlinear susceptibility. There is no second-order nonlinearity because we have assumed a system with inversion symmetry. The above iterative solutions of the equation of motion give susceptibilities as:

$$\chi^{(1)}(\omega) = -\frac{2N|\mu_{ab}|^2}{\varepsilon_0 V} \frac{1}{\hbar(\omega_{ab} - \omega) - i\hbar\gamma_{ab}} (\rho_{aa}^{(0)} - \rho_{bb}^{(0)}) \tag{1.10}$$

$$\chi^{(3)}(\omega) = -\frac{8N|\mu_{ab}|^4 \gamma_{ab} (\rho_{aa}^{(0)} - \rho_{bb}^{(0)})}{\varepsilon_0 V \gamma \left(\hbar\omega - \hbar\omega_{ab} + i\hbar\gamma_{ab} \right) \left(\hbar^2(\omega_{ab} - \omega)^2 + \hbar^2\gamma_{ab}^2 \right)} \tag{1.11}$$

Development of a light electric field under nonlinear susceptibility can be written using slowly varying envelope approximation as:

$$\frac{d}{dz}\mathbf{E}_\omega = -\frac{1}{2ik} \left(\varepsilon_0\mu_0\omega^2 + \varepsilon_0\mu_0\omega_p^2\chi^{(1)} - k^2 \right) \mathbf{E}_\omega - \frac{1}{2ik}\varepsilon_0\mu_0\omega^2\chi^{(3)}|\mathbf{E}_\omega|^2\mathbf{E}_\omega \tag{1.12}$$

where μ_0 is the vacuum permeability. This can be rewritten by separating the real and imaginary part of the susceptibility as:

$$\frac{d}{dz}\mathbf{E}_\omega = \frac{i}{2k}\varepsilon_0\mu_0\omega^2 \left[1 + \chi_R^{(1)} + \chi_R^{(3)}|\mathbf{E}_\omega|^2 - \frac{k^2}{\varepsilon_0\mu_0\omega^2} \right] \mathbf{E}_\omega - \frac{1}{2k}\mu_0\varepsilon_0\omega^2 \left(\chi_I^{(1)} + \chi_I^{(3)}|\mathbf{E}_\omega|^2 \right) \mathbf{E}_\omega \tag{1.13}$$

Suffix R denotes the real part, and I denotes the imaginary part. For this equation to hold:

$$k^2 = \varepsilon_0\mu_0\omega^2 \left(1 + \chi_R^{(1)} + \chi_R^{(3)}|\mathbf{E}_\omega|^2 \right) \tag{1.14}$$

$$\frac{d}{dz}\mathbf{E}_\omega = -\frac{\omega}{2cn} \left(\chi_I^{(1)} + \chi_I^{(3)}|\mathbf{E}_\omega|^2 \right) \mathbf{E}_\omega \tag{1.15}$$

Equation (1.14) gives the refractive index as:

$$n = \left(1 + \chi_R^{(1)} + \chi_R^{(3)}|\mathbf{E}_\omega|^2 \right)^{1/2} \tag{1.16}$$

This equation means that the refractive index changes in the optical field through a third-order nonlinear process. Equation (1.15) can be rewritten as an equation describing optical power propagation. Using:

$$\frac{d}{dz}|\mathbf{E}_\omega|^2 = \mathbf{E}_\omega^* \frac{d\mathbf{E}_\omega}{dz} + E\frac{d\mathbf{E}_\omega^*}{dz} \tag{1.17}$$

$$|\mathbf{E}_\omega|^2 = \frac{2Z_0}{n}P \tag{1.18}$$

where P is the optical power density and Z_0 is the vacuum impedance given by:

$$Z_0 = \sqrt{\frac{\mu_0}{\varepsilon_0}} \tag{1.19}$$

We can obtain the equation for the optical power density as:

$$\frac{dP}{dz} = -\frac{\alpha_0 P}{1 - \dfrac{Z_0 \chi_I^{(3)}}{2n\chi_I^{(1)}}} = -\frac{\alpha_0 P}{1 + \dfrac{P}{P_s}} \tag{1.20}$$

where α_0 is the linear absorption coefficient and is expressed as:

$$\alpha_0 = \frac{\omega}{cn}\chi_I^{(1)} = \left(\frac{2N}{V}\right)\frac{Z_0\omega\mu_{ab}^2}{n}\frac{\hbar\gamma_{ab}}{\hbar^2(\omega_{ab}-\omega)^2 + \hbar^2\gamma_{ab}^2}\Delta\rho^{(0)} \tag{1.21}$$

where $\Delta\rho^{(0)} = \rho_{bb}^{(0)} - \rho_{aa}^{(0)}$ and P_s is:

$$P_s(\omega) = -\frac{2n\chi_I^{(1)}}{Z_0\chi_I^{(3)}} = \frac{cn\varepsilon_0\gamma}{2\mu_{ab}^2\gamma_{ab}}\left(\hbar^2(\omega_{ab}-\omega)^2 + \hbar^2\gamma_{ab}^2\right) \tag{1.22}$$

Equation (1.20) means that the absorption coefficient is reduced for large optical power density and is half of the initial value for $P = P_s$. P_s is called the saturation power density and we can use this for an all-optical gate. If we introduce an intense control pulse to the two-level system, the system becomes transparent by absorption saturation. Under this condition, a weak signal light can pass through the two-level system. This is the 'on state' of the gate. When we turn off the control pulse, the system is again absorptive with a time constant of T_1, and the gate switch is in the 'off-state'. It can be seen that the absorption saturation takes place over the homogenous width of $\Delta\omega = \omega - \omega_{ab} = \hbar\gamma_{ab}$. When there is population inversion, the two-level system has optical gain, and the third order process gives the gain saturation. SOA corresponds to this case. This also can be used as an all-optical gate switch.

To examine the relationship between the response speed and the optical power density needed to saturate the two-level system, we consider the on resonant case, i.e. $\omega = \omega_{ab}$. The saturation power density is given by:

$$P_s = \frac{n\hbar^2\gamma\gamma_{ab}}{2\mu_{ab}^2 Z_0} = \frac{cn\varepsilon_0\hbar^2}{2\mu_{ab}^2 T_1 T_2} \tag{1.23}$$

The smaller T_1 and T_2 give faster response speeds while, however, smaller T_1 and T_2 give larger P_s. Large optical energy is needed for a very fast nonlinear response. This is the intrinsic limitation in using nonlinearity for all-optical signal processing devices. In the evaluation of ultrafast devices, we use short pulses. It is customary to use pulse energy rather than optical power density as a measure of device performance. The saturation pulse energy is the product of P_s, the cross section of the beam, and the pulse width. The discussion on the relationship between the response speed and optical power density (pulse energy) also holds for the refractive index, because of Kramers–Kronig relation that connects the absorption coefficient and refractive index.

More detailed analysis reveals that there are varieties of interesting effects in the optical nonlinearity. For example, if we assume pump wave ω_p and signal wave ω_s of different frequencies and consider beat frequency $2\omega_p - \omega_s$, we obtain the third-order susceptibility for nondegenerate four-wave mixing. This frequency is the beat frequency between $\omega_p - \omega_s$ and ω_p, i.e. the pump wave ω_p is scattered by the beat frequency $\omega_p - \omega_s$ to generate a new frequency $2\omega_p - \omega_s$. Detailed discussion on four-wave mixing in SOA is described in Chapter 6, which considers some other effects on the third-order nonlinear susceptibility. If we further extend the analysis to multi-level systems, we obtain expressions for multi-photon absorption and Raman scattering processes [17].

To extend the analysis from the simple two-level system to semiconductor band structures, following substitution using the wave number of electrons, $\mathbf{k}$ applies.

$$\left(\frac{2N}{V}\right) \rightarrow D\,(\mathbf{k})\,d\mathbf{k}, \text{ where } D\,(\mathbf{k}) \text{ is the density of state.}$$

Express parameters in terms of $\mathbf{k}$, for example $\hbar\omega_i \rightarrow \frac{\hbar^2 k_i^2}{2m^*}$, where m^* is the electron effective mass:

$$\rho_{aa}^{(0)}, \rho_{bb}^{(0)} \rightarrow \text{Fermi–Dirac distribution function expressed in terms of } \mathbf{k}$$

Then integrate over $\mathbf{k}$. This gives the parameters for semiconductor-based systems. It goes without saying that relationship (1.23) also holds for semiconductors.

1.5 Overview of the Devices and Their Concepts

Here we briefly review the devices described in this book in order to have an overview of their basic concepts as related to ultrafast operation. Lots of new ideas are employed and new challenges have arisen.

In Chapter 2, we describe ultrafast light sources. These are mode-lock lasers and EAM-based light sources. The mode-locked laser uses the absorption-saturation effect for mode locking. Hybrid mode locking using microwave modulation and sub-harmonic synchronous locking were employed to generate high repletion rate short pulses with small jitter. Mode-locked lasers can also be used for clock extraction from the deteriorated received signal. The 3R (retiming, retiming, regeneration) operation for a 160-Gb/s signal was demonstrated using mode-locked lasers. Also described in Chapter 2 is the EAM-based ultrafast light source. By cascading two EAMs, which are modulated by a 40-Gb/s electric signal, 3-ps width short pulses with 40 Gb/s repletion were generated. Then, a 160 Gb/s optical signal was generated by OTDM, i.e. by applying a proper time delay to four 40-Gb/s channels using space optics. An interesting point was that the CS-RZ (carrier suppressed return to zero) signal at 160 Gb/s was generated by controlling the phases of each channel by temperature. The CS-RZ modulation format is robust to nonlinear effect in the fiber, such as four-wave mixing, because of no carrier in the spectrum.

In Chapter 3, switching using a SOA (semiconductor optical amplifier) is discussed. In the SOA, population inversion is realized by current injection. When we put in an intense gate pulse, it causes a gain reduction and an associated refractive index change takes place. This is the third-order nonlinear process, and its basic principle can be understood by replacing the absorption coefficient in Equation (1.20) by the gain of SOA. A characteristic feature of this response is that it is very fast for the rise time; however, there is a slow component of the order of 1ns in the response recovery, which is the band-to-band recombination lifetime. This slow component has been the obstacle in realizing ultrafast switching devices above 100 Gb/s. Two methods are described for solving this problem. One is to use a wavelength filter to select only the very fast, blue-shifted component of the response [16, 17]. By using only the fast component, which is due to intraband electron–phonon scattering, we can perform wavelength conversion, 2R (retiming and reshaping) operation and DEMUX operation. Another method is to use the Symmetric Mach-Zehnder (SMZ) interferometer configuration. By putting SOA symmetrically at both arms of a SMZ, we can cancel out the slow response component by using gate on pulse and off pulse. Using the SMZ configuration, DEMUX operation of 640 Gb/s to 10 Gb/s was demonstrated. Error free DEMUX operations of 320 to 40 Gb/s and to 10 Gb/s were also demonstrated. Also demonstrated were the wavelength conversion and retiming and reshaping (2R) operations. It is interesting that this SMZ gate switch can be used for rather slow signal processing. Bit rate free 2R operation (2.5–42 Gb/s) was demonstrated for NRZ signals.

Chapter 4 describes a different approach to realizing ultrafast signal processing. The Uni-Traveling-Carrier Photodiode (UTC-PD), which has a very fast response with high output current, followed by a monolithically integrated traveling-wave electro-absorption modulator (TW-EAM), was used for ultrafast signal processing. In this method, very short gate pulses are converted to the electrical signal using UTC-PD, which has a 3-dB cut-off frequency of above 300 GHz. The electric signal modulates the optical signal by TW-EAM, which is monolithically integrated with the UTC-PD. Using this method, DEMUX operation of 320 Gb/s to 10 Gb/s was demonstrated. We are not bothered by the intrinsic relationship of response time and energy requirement inherent to the third-order nonlinear process. Careful design taking into account RC limit and phase matching of the traveling wave EAM modulator are crucial issues.

Chapter 5 describes the inter-sub-band transition (ISBT) gate. The ISBT gate utilizes the absorption saturation in the inter-sub-band transition in the conduction band of a very thin quantum well. To obtain a 1.55 µm transition, a deep potential well and ultra-thin well have to be employed. Material systems satisfying this are (GaN)/AlN, (CdS/ZnSe)/BeTe, and (InGaAs)/AlAs/AlAsSb material systems. Materials in parentheses are well layer materials. Response time depends on T_1 and T_2 times, which depend on parameters such as LO-Phonon energy, effective mass, and dielectric constant. Because of these parameters the (GaN)/AlN quantum well shows the fastest response, then (CdS/ZnSe)/BeTe, with the (InGaAs)/AlAs/AlAsSb system being the slowest. However, even in the slowest case, the response time is about 1 ps. Owing to the very fast response time of ISBT, we can realize a very fast absorption saturation type gate. A problem, however, is that gating energy becomes large for a fast response, as can be seen from Equation (1.23). Then in ISBT gates, much effort has been put into realizing low-gating energy despite the very fast response. In addition to the intensity gate based on absorption saturation, a new very interesting phenomenon has been found in the InGaAs/AlAs/AlAsSb ISBT gate, i.e all-optical, deep-phase modulation can be used on the loss-less TE mode probe light when the quantum well is illuminated by a TM gate pulse. This new phenomenon is quite useful, and wavelength conversion of picosecond pulses with 10 Gb/s repetition was demonstrated using this effect.

In Chapter 6, four-wave mixing (FWM) wavelength conversion using SOA is presented. Various methods of wavelength conversion are reviewed and then the chapter focuses on FWM using the semiconductor gain medium, SOA and semiconductor lasers. Various effects giving third-order nonlinear susceptibility are discussed. These are carrier density pulsation, carrier heating, and spectral hole burning. An impeding effect for the application of FWM for practical systems is the asymmetry in the wavelength conversion efficiency with respect to the pumping wavelength. The wavelength conversion efficiency from long wavelength to short wavelength is high, while the short to long is small. To solve this problem, quantum dot SOA was used, and almost symmetric wavelength conversion was demonstrated at $1.3\,\mu m$. The wavelength conversion of a 160-Gb/s signal was demonstrated at $1.55\,\mu m$, although there remained asymmetry. Also demonstrated was the use of a two-wave pumping scheme using bulk active layer SOA monolithically integrated with Mach–Zehnder interferometer configuration. This generates a replica of the signal at a different wavelength. Use of the conjugate wave for the compensation of the dispersion effect was also demonstrated using FWM.

In Chapter 7, transmission experiments performed using the devices illustrated in this book are reviewed. A 160-Gb/s based eight wavelength WDM experiment and a 320-Gb/s based ten wavelength WDM experiment were performed using mode-locked lasers (see Chapter 2) and SMZ gate switch (see Chapter 3). Another was a field experiment using an EAM-based CS-RZ light source (Chapter 2). In addition to these experiments, several recent experiments above 160 Gb/s are reviewed briefly, and the trends in very high bit rate transmission technologies are discussed. Based on these discussions and the state-of-the-art development of the devices described in this book, the technical issues to be overcome to make these devices useful and practical ones are discussed.

1.6 Summary

We saw that the development of new and higher performance devices has played a crucial role in higher bit rate systems. The increased communication capacity has brought us to a new era of the information society, where broadband Internet has had a huge impact on our daily life. We have the benefit of networks; however, we are facing the problem of increased power consumption by the network equipment. To take full advantage of the communications network we are required to establish new technologies that enable transmission of huge capacity data with minimum power consumption. One of the ways to achieve this is to introduce ultrafast, all-optical, signal processing. With ultrafast all-optical signal processing, we will be able to transmit huge amounts of data by OTDM by combining WDM and/or a multilevel scheme. To this end, semiconductor-based, ultrafast, all-optical signal processing devices are essential. There is a lot of difficulty in realizing satisfactory performances. We have discussed a third-order nonlinear process taking the simplest case, and showed that there is a restrictive relationship between the device response speed and the optical energy need for operation. In this book, readers will find a lot of new ideas and trials so far encountered in attaining our goal.

References

[1] A. Toffler and H. Toffler, *Revolutionary Wealth*, Alfred A. Knopf, New York, 2006.

[2] T. L. Friedman, *The World is Flat*, Farrar, Straus and Girnux, New York, 2005.

[3] A. R. Modarressi and S. Mohan, 'Control and management in next-generation networks: challenges and opportunities', *Communication Magazine, IEEE*, **43** (10), 94–102 (2000).

[4] T. Shimizu, D. Shirai, H. Takahashi, T. Murooka, K. Obana, Y. Tonomura, T. Inoue, T. Yamaguchi, T. Fujii, N. Ohta, S. Ono, T. Aoyama, L. Herr, N. van Osdol, X, Wang, M. D. Brown, T. A. DeFanti, R. Feld, J. Balser,

S. Morris, T. Henthorn, G. Dawe, P. Otto, and L. Smarr, 'International real-time streaming of 4K digital cinema', *Future Generation Computer Systems*, **22**, 929–939 (2006).

[5] Y. Fujita, 'Future networked broadcasting systems with ultrahigh-speed optic transmission technologies', *The Third International Symposium on Ultrafast Photonic Technologies*, Boston, August 2007.

[6] N. Nagatsu, S. Okamoto, K. Sato, 'Large scale photonic transport network design based on optical paths', *Proceedings: Global Telecommunications Conference*, Vol. 1, pp.321–327, London, 1996.

[7] M. Nakazawa, T. Yamamoto, and K. R. Tamura, '1.28 Tbit/s-70 km OTDM transmission using third- and fourth-order simultaneous dispersion compensation with a phase modulator', *Electron. Letters*, **36**, 2027–2029 (2000).

[8] T. Tokle, M. Serbay, J. B. Jensen, Y. Geng, W. Rosenkranz, and P. Jeppesen, 'Investigation of multilevel phase and amplitude modulation formats in combination with polarization multiplexing up to 240Gb/s', *IEEE Photon. Technology. Letters*, **18**, 2090–2092 (2006).

[9] K. Sekine, N. Kikuchi, S. Sasaki, S. Hayase, C. Hasegawa, and T. Sugawara, '40Gb/s, 16-ary (4 bit/symbol) optical modulation/demodulation scheme', *Electron. Letters*, **41**, 430–432 (2005).

[10] M. Nakazawa, M. Yoshida, K. Kasai, and J. Hongou, '20 Msymbol/s 64 and 128 QAM coherent optical transmission over 525 km using heterodyne detection with frequency-stabilized laser', *Electron. Letters*, **42**, 710–712 (2006).

[11] K. Kikuchi, 'Phase-diversity homodyne detection of multilevel optical modulation with digital carrier phase estimation', *IEEE Journal on Selected Topics in Quantum Electronics*, **12**, 563–570 (2006).

[12] N. J. Doran, and D. Wood, 'Nonlinear-optical loop mirror', *Optics Letters*, **13**, 56–58 (1988).

[13] T. Yamamoto, E. Yoshida, and M. Nakazawa, 'Ultrafast nonlinear optical loop mirror for demultiplexing 640 Gb/s TDM signals', *Electron. Letters*, **34**, 1013–1014 (1998).

[14] T. Sakurai, 'Ultrafast Photonic Device Technology in FST Project', *Europe–US–Japan Symposium on Ultrafast Photonic Technology*, Chiba, Japan, 2003.

[15] H. Murai, 'EA modulator based OTDM technique for 160 Gb/s optical signal transmission', *Journal of the National Institute of Information and Communication Technology*, **53**(2), 27–35 (2006).

[16] M. L. Nielsen, B. Lavigne, and B. Dagens, 'Polarity-preserving SOA-based wavelength conversion at 40 Gb/s using bandpass filtering', *Electron. Letters*, **39**, 1334–1335 (2003).

[17] Y. Liu, E. Tangdiongga, Z. Li, S. Zhang, H. de Waardt, G. D. Khoe, and H. J. S. Dorren, '80 Gb/s wavelength conversion using a semiconductor optical amplifier and an optical bandpass filter', *Electron. Letters*, **41**, 487–489 (2005).

[18] W. E. Lamb, 'Theory of an optical maser', *Phys. Rev.* **134**, A1429–A1450 (1964).

[19] Y. R. Shen, *The Principle of Nonlinear Optics*, John Wiley & Sons, Inc. Hoboken, New Jersey, 2003.

2

Light Sources

Yoh Ogawa and Hitoshi Murai

2.1 Requirement for Light Sources

As a result of the growth of broadband communications through the spread of ADSL and
FTTH, we can expect to see a rapid increase in the communication capacities of backbone
optical networks. To keep pace with such a rapid growth in data-com traffic, the number
of wavelength channels in a dense wavelength division multiplexing (DWDM) transmission
system will be continuously evolving. Upgrading a base bit-rate in a commercial system
from the current 10 Gb/s to higher is a promising way of reducing the effort involved in
managing ever more expanding optical networks. Although, today, 40-Gb/s transmission
technology has entered the stage of commercial application, it will be necessary to extend
transmission rates to 160 Gb/s or more for future ultra-high-capacity photonic networks. As
the maximum speed of available electrical-time-division-multiplexing (ETDM) circuits is currently 100 Gb/s [1–3], the so-called optical-time-division-multiplexing (OTDM) technique has
been widely used to increase transmission rates up to 160 Gb/s. The OTDM format enables
an increase in channel rate without electrical limitation, and so far, terabit-capacity OTDM
transmission has been demonstrated with a single wavelength channel [4, 5]. 160-Gb/s based
OTDM/WDM hybrid transmission experiments have been also reported, and the feasibility of
reducing wavelength channels in terabit-capacity WDM transmission has been proved [6–8].
The management of optical signal-to-noise ratio (OSNR), dispersion compensation, fiber nonlinearity, and polarization mode dispersion (PMD) tends to become more and more strict in the
160-Gb/s based transmission systems. To ease such severity in system management, forward-
error-correction (FEC) [8] and novel phase-coded signals represented by carrier-suppressed
RZ (CS-RZ) [9, 10] or RZ-differential-phase-shift-keying (RZ-DPSK) [11] have recently been
introduced to 160-Gb/s OTDM signals, and the improvement of transmission capability has
been experimentally demonstrated. Also concerning the issue of severe dispersion tolerance,
an adaptive dispersion compensation technique, able to reduce the influence of local dispersion fluctuation induced by variations in environmental conditions, has been successfully

Ultrafast All-Optical Signal Processing Devices   Edited by Hiroshi Ishikawa
© 2008 John Wiley & Sons, Ltd

developed [12]. In addition, the advance of adaptive PMD mitigation techniques [13] and optical 3R regenerators [14] strongly supports the potential to realize future 160-Gb/s based optical links.

The OTDM technique performs time-division multiplexing and demultiplexing of two or more optical data sequences as shown Figure 2.1. In principle, it enables an increase in channel rate without electrical limitation and can generate ultra-high-speed signals at a single wavelength, the speed depending on the extent of multiplexing. Important elemental techniques for OTDM signal generation include ultra-short optical pulse generation, which prevents interference between multiplexed optical data; and optical time division multiplexing, which bit interleaves individually data-modulated optical signals. For signal reception, OTDM also requires a clock extraction technique that extracts the system clock synchronized with the OTDM signal, and optical demultiplexing, which separates multiplexed channels without crosstalk.

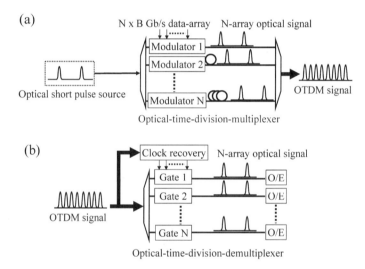

Figure 2.1   Schematic image of OTDM technology showing (a) OTDM transmitter, (b) OTDM receiver

In this section, we describe the main criteria required for optical short-pulse source and optical-time-division-multiplexer.

2.1.1 Optical Short Pulse Source

In an optical short pulse source, the arrival time fluctuation of signal pulses, i.e. a timing jitter, is one of the dominant limiting factors for signal quality. Particularly, at a bit rate beyond 100 Gb/s, the signal quality becomes more and more sensitive to the timing jitter. The acceptable value of timing jitter can be simply estimated by a simple model introduced in References [15–17], for example. Assuming that the detection error is induced only by the random timing jitter with Gaussian distribution and that the mark ratio of signal is 50 %, the error probability of a detected signal is described using the complementary error function:

$$P_{err}\left(\frac{Tw}{2\sqrt{2}\sigma}\right) = \frac{1}{\sqrt{\pi}} \int_{Tw/(2\sqrt{2}\sigma)}^{\infty} e^{-\tau^2} d\tau$$

$$= \frac{1}{2} erfc\left(\frac{Tw}{2\sqrt{2}\sigma}\right) \tag{2.1}$$

where Tw and σ denote the detection window and RMS timing jitter. $Tw/2\sigma$ can be regarded as the signal-to-noise ratio of a detected signal, that is, it corresponds to the so-called Q-factor and is expressed by:

$$Q = \frac{Tw}{2\sigma} \tag{2.2}$$

Although Equations (2.1) and (2.2) are not rigorous expressions for the error probability induced by random timing jitter, they enable us to estimate roughly the acceptable timing jitter. For example, consider the RMS jitter required for optical pulse train in a 160-Gb/s system. When setting a target Q-factor greater than 20 at a transmitter and assuming a detection window of 6 ps at the receiver side, the random jitter must be less than 150 fs. In a typical OTDM system, the optically multiplexed signal is to be down-converted to a lower bit-rate signal as shown in Figure 2.1(b). In this case, the time window for square-law detection is significantly widened, so that the signal quality becomes less sensitive to the timing jitter. For instance, since the detection window of an optical signal demultiplexed to 40 Gb/s becomes four times that of a 160-Gb/s signal, the limitation of the random timing fluctuation is simply relaxed to 600 fs. However, the quality of the optically demultiplexing signal strongly depends not only on signal timing jitter, but also on the shape and the timing jitter of the gate window. Hence, a timing jitter below the acceptable value given by Equation (2.2) would be required for an optical pulse source.

When focusing on an OTDM format, the optical pulse width should be carefully adjusted so that the optical pulse trains can be multiplexed without coherent interference between them. The optically multiplexed signal with a duty cycle of less than 50 % requests the optical pulse with a duty cycle below $(50/N)$ % in the $N \times B$ Gb/s OTDM system shown in Figure 2.1(a). The extinction ratio of optical pulse train will also be important, because its decrease results in a degradation of eye-opening and bit-by-bit pulse-peak fluctuation induced by coherent interference between the multiplexed channels. Here, we consider an OTDM system with a multiplexing number of N and a signal pulse with a extinction ration of ER. The degree of eye-opening in an N-multiplexed signal is strongly dependent on the value of N in addition to the pulse extinction ratio. Figure 2.2 represents an example of an OTDM signal assuming that $N = 4$ and $ER = 20$ dB. The optical eye-diagram shows that an insufficient extinction ratio results in significant intensity fluctuation in both 'mark' and 'space', and mark/space extinction; the $ER_{m/s}$ ratio is reduced by a factor of $(N-1)^2$ compared with the pulse extinction ratio, ER. In accordance with our numerical analysis, shown in Figure 2.3, a mark/space extinction ratio greater than 22 dB in the OTDM signal is required in order to get an eye-opening penalty of less than 0.5 dB, and such a criterion requires that the extinction ratio of the input pulse train must be greater than 32 dB for $N = 4$, and 38 dB for $N = 8$. That is, the requirement on

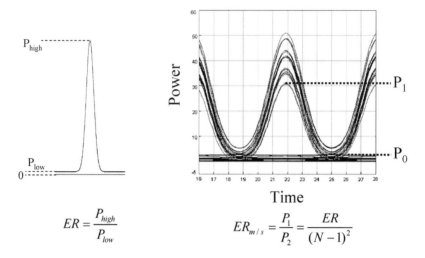

$$ER = \frac{P_{high}}{P_{low}} \qquad\qquad ER_{m/s} = \frac{P_1}{P_2} = \frac{ER}{(N-1)^2}$$

Figure 2.2 The impact of a low extinction ratio on an OTDM signal. The optical eye diagram corresponds to an OTDM signal with $N = 4$ and ER $= 20\,$dB

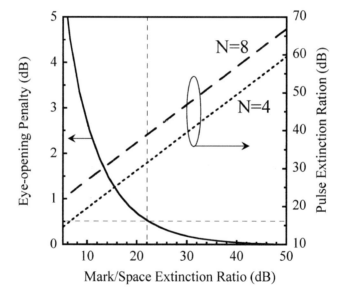

Figure 2.3 Eye-opening penalty and extinction ratio; ER required for input pulse train (dotted line, $N = 4$; dashed line, $N = 8$), plotted as a function of mark/space extinction ratio; $ER_{m/s}$

the extinction ratio of an optical pulse source becomes more stringent with the increase in multiplexing channels.

As mentioned above, the deterioration in the extinction ratio also induces bit-by-bit pulse-peak fluctuation, depending on the relative phase between adjacent bits. The worst case, where the interference significantly impairs signal quality, is when the relative phase of a particular channel has shifted by π against the other multiplexed channels (e.g. '$\ldots 0\,\pi\,0\,0\,0\ldots$'). As the pulse extinction ratio decreases, the pulse-peak of out-of-phase, channel should be depressed with a rate of $(\sqrt{ER} - N + 1)^2/(\sqrt{ER} + N - 3)^2$. In case of an extinction ratio of 30 dB and $N = 4$, the pulse-peak is depressed by about 20 %, and as a result it causes an additional power penalty of 1 dB in the depressed channel. In order to suppress the additional penalty below 0.5 dB ($< 10\,\%$ peak depression), an extinction ratio above 40 dB is imposed on the optical pulse train. In addition to such requirements, the optical pulse source has to be highly stable and compact in terms of practicality. Semiconductor-based mode-locked laser diode (MLLD) [18–21] is known to be a suitable pulse source for meeting such requirements, and the applicability to OTDM signals has been widely investigated [22–24]. The detail of the MLLD is described in next section. As a simpler alternative, the optical pulse generation utilizing InGaAsP-based multiple-quantum well (MQW) EA modulator was also widely employed [25–27]. The EA modulator is a simple gate device utilizing optical absorption change depending on the applied electric field. Such an effect is known as the Franz–Keldysh effect in bulk waveguide and also the quantum confined Stark effect (QCSE) [28] in MQW waveguide. A unique characteristic of such kinds of effect is that the optical absorption changes nonlinearly with applied field, and thanks to the nonlinear response, the EA modulator can act as an optical gate with a very narrow window. Consequently, it enables generation of optical short pulses in a different manner from MLLD. In Section 2.3, we describe the various applications of EA modulators for processing ultra high-speed optical signals.

2.1.2 Optical Time Division Multiplexer

The optical time division multiplexer is another essential element for ultra high-speed signal source. In most ultra high-speed OTDM transmission experiments, the passive-delay-line optical circuit has been used as the optical multiplexer, since it is convenient to make a tentative ultra high-speed signal, such as a 160-Gb/s signal. From a practical point of view, however, the optical multiplexing circuit should integrate optical modulators for data encoding of all multiplexed channels, optical delay lines for bit-interleaving, and power splitter/combiners within a compact package. The number of multiplexed channels is one of the most important issues to be considered here, since increasing the multiplexing channels results not only in difficulty in its design, but also in more severe restriction of the extinction ratio imposed on the optical short pulse source. Moreover, the insertion loss of the optical multiplexer could be enhanced by increasing the multiplexing channels. Thus, a small number of multiplexing is preferable. The minimum degree of multiplexing is mainly limited by the available capability of optical encoding that is implemented in the optical multiplexer. The current highest modulation bandwidth for commercially available devices would be restricted below about 50 Gb/s, regardless of electronic or optical components, although state-of-the-art technology might make the 100-Gb/s class ETDM modulation realistic [2, 29] in the near future. The OTDM system with multiplexing 40-Gb/s optical channels is the current best solution for generating an ultra high-speed signal of more than 100 Gb/s.

The major issue in the realization of an authentic optical multiplexer, which integrates the functions of all independent encoding, arises from the difficulty in design of high-speed microwave circuits. Regarding a two-multiplexing system, it is relatively easy to integrate two optical modulators monolithically in an optical multiplexer. For example, optical time-division multiplexers integrating two lithium niobate (LiNbO$_3$)-based Mach–Zehnder modulators have been reported [30, 31]. The monolithic integration of optical modulators quickly becomes difficult as soon as the degree of multiplexing increases to $N = 4$ or more, due to electromagnetic interference. To overcome such a limitation, utilizing all optical signal processing could be a promising option. Planner light wave circuit (PLC)-based, all-optical, multiplexer integrated, periodically poled lithium niobate (PPLN) has been reported [32], and all-channel independent encoding of 4×40-Gb/s multiplexing channels has been successfully achieved. As another useful approach to realizing an authentic multiplexer, namely a free-space integration of lens-coupled modular-type EA modulators, has also been proposed [10]. Regarding the free-space, optics-based, optical multiplexer, this will be described in detail in Section 2.3 of this chapter.

2.2 Mode-locked Laser Diodes

The operational principle of mode locking is the synchronous modulation of an optical pulse that goes around the resonator. Modulating a loss of a mode locker inserted in the cavity at the round trip frequency makes the optical pulse shorten every pass through the mode locker. As the result of the balance between the widening and shortening, pulse oscillation occurs. Another explanation in the frequency domain is phase synchronization between longitudinal modes through the modulation side band. Modulation side bands are created in the vicinity of the other longitudinal modes of both sides because the mode separation corresponds to the round trip frequency. Every longitudinal mode is pulling to the modulation side band, and then the phases of all modes are synchronized.

An MLLD is a promising device as a light source for all-optical processing systems because it has excellent features of a semiconductor laser and mode-locking technique. There are two types of MLLD differentiated by the modulation methods: active mode locking for electrical modulation, and passive mode locking for optical modulation. In an active MLLD, the mode-locking process is forcibly induced by external modulation and an output pulse train is synchronized with the external signal. On the other hand, a passive MLLD needs no external modulation source for pulse generation and the characteristics do not suffer from electrical limitation, whereas the repetition frequency is not synchronized with external electrical signals. In order to solve the above problem during practical use, hybrid mode locking and optical synchronous mode locking have been developed.

From the point of view of cavity configuration, MLLDs can be categorized into two types, namely external cavity [33, 34] and monolithic types. An external-cavity MLLD has the advantages that usable functions such as a wavelength tuning and repetition frequency tuning can be added by inserting optical components into the cavity. A monolithic MLLD, in contrast, has the advantage of a smaller size and mechanical stability as compared with an external cavity MLLD. These features are promising for practical system applications.

In this section, we describe the monolithic type of active and passive MLLDs for optical communications systems. Furthermore, the synchronization and stabilization techniques for a passive MLLD and the application for clock extraction are also described.

2.2.1 Active Mode Locking

Active mode locking arises from the electrical modulation of gain or loss in a laser cavity at a frequency near the round-trip frequency. Figure 2.4 shows a schematic cross section and

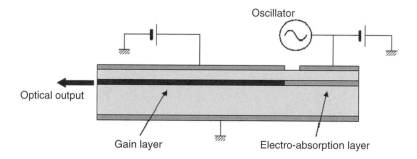

Figure 2.4   Schematic cross-section and driving method of an active MLLD

driving method of an active MLLD. In this case, a laser diode is integrated with an electro-absorption (EA) modulator in the cavity [21, 35, 36]. Another feature of active mode locking is that it modulates directly to a part of the gain section without an EA modulator [37].

The round-trip frequency f_r of a laser diode is given in the following expression:

$$f_r = \frac{c}{2n_g L} \tag{2.3}$$

where c is the velocity of light in a vacuum, n_g is the refractive index for the group velocity in the cavity, and L is the cavity length. For example, the round-trip frequency is about 43 GHz when $L = 1$ mm and $n_g = 3.5$.

When a RF electric field at the frequency of f_M near f_r is applied to the electro-absorption layer, the absorption coefficient α is described by:

$$\alpha = \alpha_0(1 + M(1 - cos2\pi f_M t)) \tag{2.4}$$

where α_0 is a constant, and M is the modulation coefficient, and f_M is the modulation frequency. The repetition frequency is equal to f_M, and close to the round-trip frequency. According to Haus's theory [38], the output pulse shape of active mode locking becomes a Gaussian waveform, and the pulse width is given by

$$\tau = \frac{\sqrt{2\sqrt{2}ln2}}{\pi} \frac{1}{\sqrt[4]{M/g}\sqrt{f_M f_L}} \tag{2.5}$$

where g is the gain of the peak, f_L is the bandwidth of the gain, and the dispersion of the resonator is disregarded. It is an important relationship that the pulse width is inversely proportional to the fourth root of the modulation index M.

Figure 2.5 shows the structure of a monolithic type of active MLLD device [21]. The active layer of the gain section consists of InGaAsP/InGaAsP multiple quantum wells, and the EA, waveguide and the DBR sections consist of a 1.465-μm band gap wavelength InGaAsP bulk layer. A ridge-waveguide structure buried with polyimide is used for lateral confinement. Figure 2.6 shows (a) a typical SHG correlation trace, and (b) a typical optical spectrum when

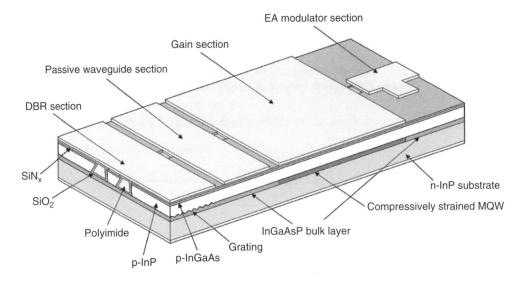

Figure 2.5 Device structure for a monolithic type of active MLLD

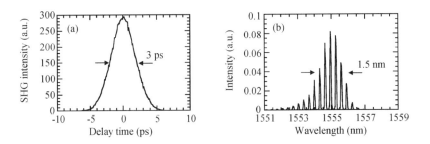

Figure 2.6 (a) SHG correlation trace and (b) optical spectrum of an active MLLD

the gain current is 50 mA, the EA bias voltage is 0.2 V, RF power is 23 dBm, and operational frequency is 39.81312 GHz, which is the clock frequency of the standard 40-Gb/s optical communication systems. The pulse width was estimated to be 3 ps, assuming a Gaussian waveform. The time–bandwidth product is 0.46, close to the transform limited value of 0.44 for a Gaussian waveform. Figure 2.7 shows the dependence of the pulse width on RF power. In linear approximation, M is considered to be proportional to the RF voltage and the RF voltage is proportional to the square root of RF power. So Equation (2.3) coincides well with the experimental results in Figure 2.7. Figure 2.8 shows the measurement results of the pulse width when the RF frequency is changed around the round-trip frequency. In the case of active mode locking, the locking range was dominated by the degradation of the RIN value for the increase of the loss, by shifting a peak of the transmissivity from the peak of the pulse. When we define the locking range as a frequency region in which the degradation of the RIN value was within 6 dB (corresponding to a doubling of the amplitude noise), the locking range was from 39 GHz to 40.9 GHz.

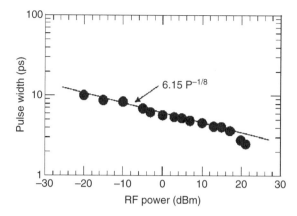

Figure 2.7 Dependence of pulse width on RF power of an active MLLD

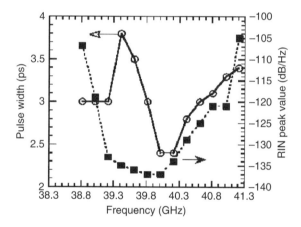

Figure 2.8 Dependence of RIN peak value and pulse width on the RF frequency of an active MLLD

2.2.2 Passive Mode Locking

Figure 2.9 shows a schematic cross section and driving method of a passive MLLD. In the case of passive mode locking, a saturable absorber (SA) is substituted for the EA modulator of an active MLLD. A multiple quantum well (MQW) structure of same material as the gain section is ordinarily used as an SA [39]. When currents are injected into a gain section and an SA is reverse biased, passive mode locking occurs. Note that the pulse generation can only be obtained by DC bias without high-speed electronics. Generally, an SA is placed at a resonator edge where a light pulse arrives once in every round-trip. In contrast, a configuration where an SA is located at the center of a laser cavity is called a colliding pulse mode-locked laser diode (CPM-LD) [40]. The CPM can reduce the absorption saturation energy equivalently by coherent interaction between incoming optical pulses from both sides of the gain section at the SA.

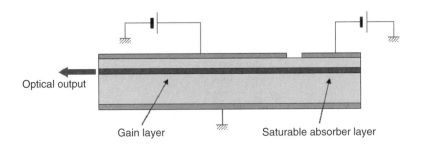

Figure 2.9 Schematic cross-section and driving method of a passive MLLD

The mode-locked frequency f_{ML} for passive mode locking is N times round-trip frequency (where N is an integer), see Equation (2.3). It is called fundamental mode locking when $N = 1$ and Nth order harmonic mode locking when $N = 2$ [41]. Figure 2.9 illustrates a case where an SA is placed in the resonator edge and $N = 1$, and a case where CPM configuration usually becomes $N = 2$. In addition, higher order harmonic mode locking can be performed by putting an SA at a suitable position utilizing a CPM effect [42].

According to Haus's theory [38, 43], the pulse shape of passive mode locking becomes a sech^2 waveform, and the pulse width is given by:

$$\tau = \frac{2ln(3 + \sqrt{8})}{\pi} \frac{1}{f_c\sqrt{q^{(i)}}\left(\dfrac{E_p}{E_A}\right)\sqrt{1 - \dfrac{g^{(i)}}{q^{(i)}}\left(\dfrac{E_A}{E_L}\right)^2}} \tag{2.6}$$

where E_p is the pulse energy, E_A is the absorption saturation energy, E_L is the gain saturation energy, $q^{(i)}$ is the saturable absorption normalized by the cavity loss before passage of the pulse, $g^{(i)}$ is gain normalized by the cavity loss before passage of the pulse, and f_c is the gain bandwidth. The pulse width is narrower, as the pulse energy is larger, the saturation energy is smaller, and nonlinear absorption is larger. This means that passive mode-locking characteristics depend on gain and absorption modulation depth and on the intracavity pulse energy. The next condition must be satisfied in order to have a real solution for Equation (2.6):

$$\frac{g^{(i)}}{q^{(i)}}\left(\frac{E_A}{E_L}\right)^2 < 1 \tag{2.7}$$

This is the necessary condition for the absorption saturation energy to be smaller than the gain saturation energy.

Figure 2.10 shows the schematic structure of a monolithic type of passive MLLD The SA and the gain section consist of InGaAsP/InGaAsP strained multiple quantum wells, and the passive waveguide and DBR section consist of InGaAsP bulk layers of 1.3 μm band-gap wavelength. Figure 2.11 shows measurement results for the SHG correlation trace and the optical spectrum of the output pulse. The repetition frequency was about 40 GHz, corresponding to the round-trip frequency estimated from a cavity length of about 1.1 mm of the MLLD used. The correlation trace and the envelope of the optical spectrum agree well with a sech^2 waveform; the pulse width was 4.3 ps and spectral width was 0.62 nm. The time–bandwidth product is 0.33, very close to the transform limited value of 0.315.

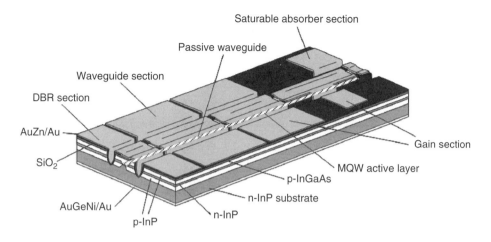

Figure 2.10 Device structure for monolithic type of passive MLLD

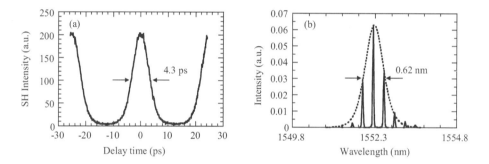

Figure 2.11 (a) SHG correlation trace, and (b) optical spectrum of a passive MLLD

2.2.3 Hybrid Mode Locking

Figure 2.12 shows the configuration of a hybrid mode-locking. RF signal when a frequency around the round-trip frequency of the passive MLLD is applied to the SA [44]. The electrical spectra of the hybrid MLLD are shown in Figure 2.13. The electrical spectrum that was broad with a Lorenzian shape for the passive mode-locking operation ($P_{rf} = 0$) drastically changed to sharp once an RF signal was applied to the SA ($P_{rf} = 16$ dBm), so it is hybrid mode locking. This indicates that the frequency of passive mode locking was synchronized with the frequency of the external signal applied to the SA.

The jitter is generally found by integrating the noise component of a measured electrical spectrum. The next relationship is formed as $L(f)$ is the SSB noise (Single Side Band Noise) level relative to the carrier per 1 Hz [45, 46].

$$\left(\frac{\Delta E}{E}\right)^2 + (2\pi n)^2 \left(\frac{\Delta t}{T}\right)^2 = 2 \int_{f_{low}}^{f_{high}} L(f) df \tag{2.8}$$

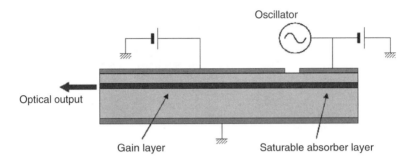

Figure 2.12   Schematic cross-section and driving method of a hybrid MLLD

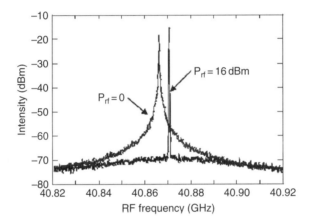

Figure 2.13   The electrical spectrum of MLLD with RF power of 16 dBm (hybrid mode locking) and without RF power (passive mode locking)

Where ΔE is the r.m.s. amplitude jitter, Δt is the r.m.s. timing jitter, E is the average pulse energy, T is the repetition time, f is the off-set frequency from the carrier frequency, f_{low} and f_{high} are the lower and upper limits of the integral range, and n is the harmonic number. The timing jitter and amplitude jitter can be divided by measuring the dependence on harmonic number.

The timing jitter was estimated at about 4 ps for passive mode-locking operation and 0.18 ps for hybrid mode-locking operation by integrating the electrical spectrum in Figure 2.13 from 100 Hz to 10 MHz. Where the first term on the left-hand side in Equation (2.8) is assumed to be 0, it is supposed that the amplitude jitter is negligible. The timing jitter for hybrid mode-locking operation was almost same as that of the synthesizer that was used in the experiment.

In the case of hybrid mode locking, the pulse characteristics including pulse shape and pulse width, are determined by the dynamics of gain and absorber saturation, and an applied RF signal almost does not affect it. The degradation of the timing jitter is a dominant factor in restricting the locking range when the frequency of the RF signal shifts from that of passive mode locking. Figure 2.14 shows the dependence of timing jitter and pulse width on the injected

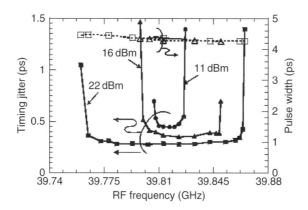

Figure 2.14   Frequency locking characteristics of the hybrid MLLD as parameter of RF power

RF frequency as a parameter of the RF power. The change in pulse width was less than 0.2 ps from 39.76 GHz to 39.86 GHz. The timing jitter decreased and the locking range broadened as the injected RF power was increased. In this example, the locking range of 100 MHz was obtained at an RF power of 22 dBm.

2.2.4 Optical Synchronous Mode Locking

Figure 2.15 shows the configuration of optical synchronous mode locking. A stabilized optical pulse train is injected into the passive MLLD as a slave laser [47]. The injected optical pulse is absorbed at the SA, and then the absorption coefficient is modulated by absorption saturation. In optical synchronous mode locking, the configuration is slightly complicated, but the locking range is broad compared with hybrid mode locking because the modulation of the SA is efficient. The most important advantage is the availability of subharmonic synchronous mode locking (SSML) [48–50] that injects the optical pulse train with a repetition frequency corresponding to the subharmonic frequency of the passive mode locking. The injected optical pulse goes around a resonator and modulates the SA at a frequency the same as the repetition of the MLLD. When the mode-locking frequency is extremely high (> 100 GHz), SSML is the most promising way of stabilizing the mode-locked laser, because this is achieved using low frequency master laser pulses and application is not limited by the bandwidth of the drive electronics.

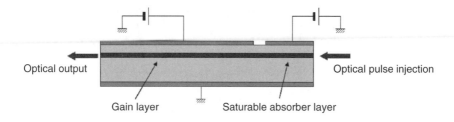

Figure 2.15   Schematic cross-section and driving method of an optical synchronous MLLD

Figure 2.16 shows an experimental setup for the SSML to stabilize a 160-GHz CPM-LD. A CPM-LD of device length 540 μm was used as the slave laser. The repetition frequency for the passive mode locking was about 158 GHz, and the pulse width was 1.68 ps. The master laser was a mode-locked erbium-doped fiber ring laser (ML-EDFL). The center wavelength and pulse width were 1551 nm and 1.94 ps, respectively. The repetition frequency was 9.873 GHz, corresponding to a 16th subharmonic of the CPM-LD's mode-locking frequency. The ML-EDFL output was compressed to 0.59 ps by adiabatic soliton compression in a dispersion decreasing fiber and then coupled to the CPM-LD. The timing jitter was 0.21 ps when the integration area ranged from 100 Hz to 100 MHz.

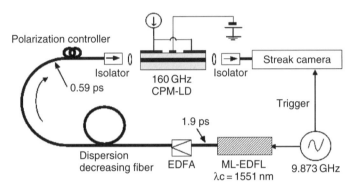

Figure 2.16   Experimental setup for an SSML

Figure 2.17 shows the changes in the synchroscan streak camera traces of CPM-LD outputs with and without master laser pulse injection [51]. The injection power was about 57 μW. No clear pulse train was observed without master laser pulse injection, although the optical pulse

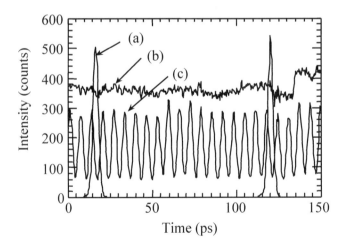

Figure 2.17   Synchroscan streak camera traces of (a) the master laser pulses, (b) the CPM-LD output without injection, and (c) with injection power of 57 μW [51]). (Reproduced by permission of ©2000 IEICE)

was generating by passive mode locking. This is because the optical pulses from the CPM-LD were not synchronized with the synchroscan streak camera. On the other hand, a clear pulse train with a repetition frequency corresponding to 16-times that of the master laser pulse's repetition frequency was observed under SSML operation when the pulse train of the master laser was injected. This indicates that the timing of the pulse generation of the CPM-LD was controlled by the injected master laser pulses, and that the CPM-LD output was synchronized to the master laser. In addition, the timing jitter of the CPM-LD was estimated by measuring the timing jitter of a beat signal between the CPM-LD output and the reference pulses at around 10 GHz. As the result, the timing jitter of 2.2 ps was reduced to 0.26 ps by SSML operation.

2.2.5 Application for Clock Extraction

Clock extraction is an essential function in optical communications systems. It is usually realized by electrical means after the optical signal is converted to an electrical signal. However, this method cannot operate at high bit rates of more than 100 Gb/s in OTDM systems because of electrical speed limitation. A passive MLLD operates as an optical clock extraction device by injection locking so that the frequency is synchronized with that of the injected external signal. In this case, the system of clock extraction is achieved by using a single chip and is very effective because it operates with all optical processes and suffers from no electrical bandwidth limitation.

Figure 2.18 shows an experimental setup for clock extraction. The clock extraction device was a 160-GHz CPM-LD as used in the experiment for the SSML in Section 2.2.4. Signal pulses were generated from a 40-GHz active MLLD. The pulse repetition frequency, pulse width, and center wavelength were 39.67137 GHz, 2.3 ps, and 1553 nm, respectively. The 40-GHz MLLD output was encoded at 39.67137 Gb/s using an EA modulator driven by a 40-Gb/s pulse pattern generator. The EA modulator output was multiplexed to 158.68548 Gb/s in the time domain using a PLC-based OTDM multiplexer after amplification by EDFA. The PLC output was coupled to the 160-GHz CPM-LD that was used for the SSML experiment. The polarization state was adjusted so as to be in a transverse-electric mode relative to the 160-GHz MLLD using a polarization controller.

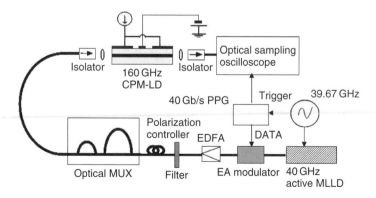

Figure 2.18 Experimental setup for clock extraction

Clock extraction from distorted signals having large amplitude and timing fluctuation is of interest for practical applications. Figure 2.19 shows optical sampling oscilloscope traces of (a) input data and (b) output clocks when the pulse width of the input data was intentionally broadened to 4.7 ps using an optical fiber [52]. The input data had unequal peak intensities and unequal time intervals and the eye opening was insufficient. This is because neighboring pulses interfered with each other due to a high duty ratio, and the phase relationship between the neighboring pulses not being identical. In contrast, the output clock was very clear. It had equal peak intensities and equal time intervals corresponding to a data bit rate of incoming data signals. This implies that retiming and reshaping functions were realized in the CPM-LD due to a cavity resonance effect and the stable clock extraction was expected despite poor quality of input signals. Such functions are also desirable for all-optical 3R regenerator [53]. All-optical 3R can be completed if the obtained clock pulses are modulated by incoming data pulses using an all-optical switching device such as that described in this book.

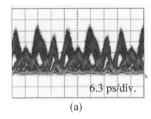 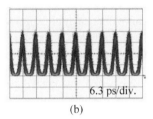

| (a) | (b) |

Figure 2.19 Optical sampling oscilloscope traces of (a) input data, and (b) output clock from the CPM-LD [52]. (Reproduced by permission of IEEE ©2004)

2.3 Electro-absorption Modulator Based Signal Source

2.3.1 Overview of Electro-absorption Modulator

An electro-absorption modulator (EAM) is a well-known optical intensity modulator as well as the LiNbO$_3$-based Mach–Zehnder modulator. The performance of EAMs has been rapidly improved since the processing technology of III-V semiconductor multiple-quantum-well (MQW) waveguide was established in the 1990s, and the EAM using a quantum confined Stark effect (QCSE) [28] today has come to be an essential device for different kinds of high-speed optical signal processing. The beneficial points of EAM are that its size is compact and also that it can operate at low voltages of 2 – 3 V, as well as affording easy integration with the light source [54–56].

Figure 2.20 illustrates the device structure of an EAM [57]. An InGaAsP-based MQW optical absorption layer is epitaxially grown on the n-type InP substrate by metal organic chemical vapor deposition (MOCVD). Both edges of the device are coated with anti-reflection film to suppress undesirable optical reflections. The modulation bandwidth of an EA device is generally restricted by the chip capacitance, and in order to reduce the chip capacitance for high-speed operation, a ridge structure for the optical waveguide, which effectively reduces the junction capacitance, is widely adopted. For further reduction of chip capacitance, the size of the pad electrodes, which induce parasitic capacitance, is minimized and a polyimide layer having a low dielectric constant is inserted between optical waveguide layer and the pad electrodes.

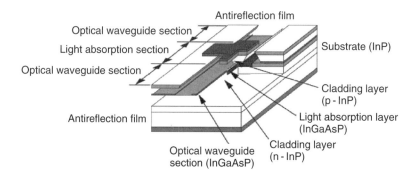

Figure 2.20   Device structure of an MQW-EAM. (Reproduced by permission of © 2007 IEEE)

The length and thickness of the optical absorption layer are optimized for maintaining low capacitance and sufficient optical absorption efficiency. The length of the EA section is around 100 μm, and including low-loss optical waveguides on either side of the absorption section, it results in a total device length of about 250 μm. The module structure of EA modulator is shown in Figure 2.21. The optical beam couples to the EA waveguide using aspherical lenses with a high coupling efficiency. A high-speed electrical signal of the order of 40 Gb/s propagates a 50-Ω microstrip line and is terminated at a resistance of 50 Ω. The performance of electro-absorption is stabilized by a constant control of operating temperature. All the components are housed inside a hermetically sealed metal package.

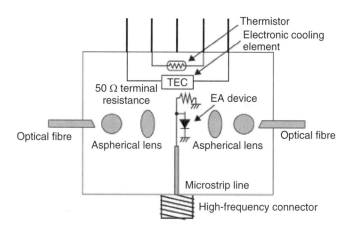

Figure 2.21   Module structure of an EA modulator. (Reproduced by permission of © 2007 IEEE)

Figure 2.22 illustrates the electro-absorption property of a typical EAM designed for a C-band transmitter. Optical loss increases nonlinearly with applied bias voltage to EAM. To obtain an extinction ratio of more than 20 dB only −4 V of bias voltage is needed. Frequency characteristics of EAM are described by E/O response and electrical reflection (S11), as represented in Figure 2.23. This shows that potentially EAM has a modulation bandwidth of more than 50 GHz. More ultra high-speed operation beyond 100 GHz is also made available by adopting

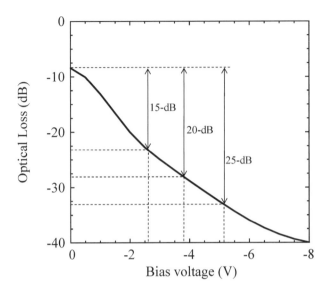

Figure 2.22 An example of the optical absorption characteristics of an EAM. CW light was input at a wavelength of 1550 nm. The operating temperature was 25 °C

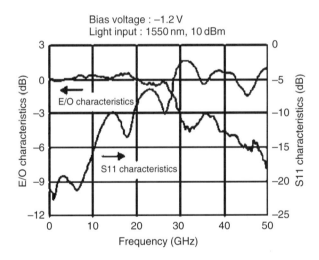

Figure 2.23 Frequency response of EAM

a traveling wave electrode [58] as well as a LiNbO3-based optical modulator. Figure 2.24 illustrates the typical chirp characteristics of EAM. The chirp characteristic is generally assessed by a figure called the 'α-parameter' [59] and desirably the α-parameter should be close to zero. As the graph shows, in this case, the α-parameter reaches 0 at a bias voltage near −1 V, which suggests that if the modulation signal is supplied at a bias voltage in the region of −1 V, then the device will operate in the low-α region, and hence low wavelength chirp can be expected.

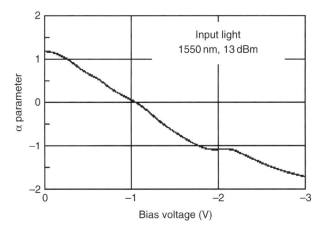

Figure 2.24 Chirp characteristic of EAM

2.3.2 Optical Short Pulse Generation Using EAM

EAM is convenient for generating optical short pulses due to its nonlinear optical absorption property against bias voltage. Figure 2.25 illustrates the fundamental configuration of an external modulation optical pulse source using EAM. In this scheme, continuous wave (CW) light from a single-mode laser diode (LD), such as a distributed feed back (DFB) LD, is modulated by sinusoidally driven EAM. For the simple understanding of optical pulse generation using EAM-based external modulation, let us assume that the optical loss of EAM is a function of bias voltage and its dependence is given by:

$$EX(V) = C\exp(-K|V|) \tag{2.9}$$

where C, K, and V are the initial loss at $V = 0$, a slope factor to determine loss curve, and bias voltage, respectively. When a sinusoidal modulation signal is applied to EAM as shown in Figure 2.25, the applied bias is expressed by:

$$V(t) = V_b + \frac{V_{pp}}{2}\cos(2\pi\Omega t)(V(t) < 0) \tag{2.10}$$

where V_b, V_{pp}, and Ω are DC bias (< 0), amplitude of sinusoidal modulation(> 0), and modulation frequency, respectively. Expanding Equation (2.10) in a Taylor series the second order and assuming $\exp\{-K|V_b - V_{pp}/2|\} \sim 0$, the transfer function of sinusoidally-driven EAM is approximates to:

$$T(t) \approx C'\exp\left(-\pi^2 KV_{pp}\Omega^2 t^2\right)$$

$$C' = C\exp\left\{-K\left|V_b + \frac{V_{pp}}{2}\right|\right\} \tag{2.11}$$

The transfer function, Equation (2.11), corresponds to a Gaussian function, and indicates that CW light input into a sinusoidally driven EAM is converted into the optical pulse close to

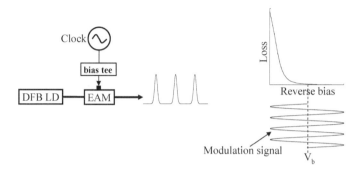

Figure 2.25 Schematic image of an EAM-based optical pulse source and the principle of optical pulse generation

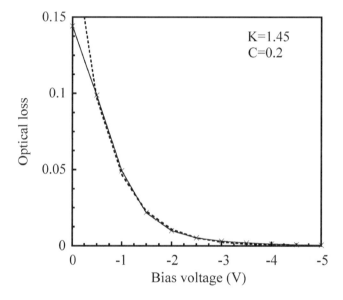

Figure 2.26 Linear plot of Figure 2.22. The dashed line corresponds to a simple fitting function as given by Equation (2.3.1)

Gaussian function. Figure 2.26 illustrates the linear plot of Figure 2.22, and the dashed line is a fitting curve drawn by using Equation (2.11) with $K = 1.45$ and $C = 0.2$. The intuitive but useful expression of Equation (2.9) exhibits good agreement with experimental results at a bias of less than -0.5 V. By use of the fitting parameter K, the pulse width of the output pulse train can be roughly estimated. Figure 2.27 represents an output pulse width estimated using Equation (2.11). Symbols plotted in Figure 2.27 are the experimental results of 20-GHz optical pulse generation. The pulse widths were measured by an autocorrelation. Under the conditions where $|V_b - V_{pp}/2| > 0.5$ V, it can be seen that the estimated pulse width using Equation (2.11) is consistent with experimental result. The simple expression in Equation (2.11) also suggests that the pulse width of the output pulse train is to be shortened proportionally to the modulation frequency. In accordance with Equation (2.11), a pulse width shorter than 5 ps should be obtained at a modulation frequency of 40 GHz.

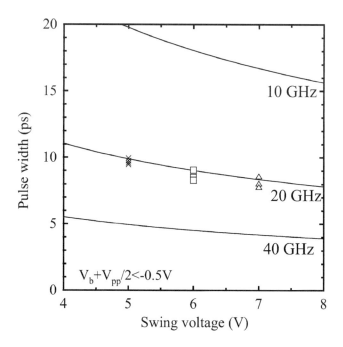

Figure 2.27 Roughly estimated output pulse width (FWHM) using Equation (2.3.3). Symbols are experimental results taken at a modulation frequency of 20 GHz

Due to the electrical limitations of the EAM module, however, the sinusoidal modulation signal supplied to an EAM tends to decrease with increasing modulation frequency. Hence, the modulation frequency response of the EAM plays an important role in ultra short pulse generation. Figure 2.28 represents modulation E/O response and return loss (S11) characteristic of the EA modulator specialized for short-pulse modulation (SPM-EAM) at 40 GHz. The SPM-EAM is unique in that it is tuned to minimize return loss (S11) of modulation signal at around 40 GHz, using a built-in impedance-matching circuit [60,61]. Such a special frequency response enhances the modulation efficiency of a 40-GHz-driven EA modulator, and enables generation of an optical pulse train shorter than 5 ps. As a 160-Gb/s OTDM signal, however, requires a pulse width of less than 3.2 ps, the SPM-EA modulator is not available yet for this case.

For the sake of further compressing the optical pulse, it is very useful to concatenate two SPM-EAMs [61–64], as shown in Figure 2.29. Two SPM-EAMs are serially connected through an erbium-doped fiber amplifier (EDFA) and are driven by a 40 (39.81312)-GHz clock signal separately. At the first stage EAM, CW light is converted into a 40-GHz optical pulse train, and after compensating for optical loss by EDFA the second EAM is converted again at 40 GHz to constrict the pulse width. According to Equation (2.11), the transfer function of the cascaded EAMs takes the form of $T(t)^2$ under the assumption that the two EAMs have the same modulation response. Thus, the gate width should be reduced by a factor of $\sqrt{2}$.

Figure 2.30 shows the pulse width variation measured as a function of bias voltage. Triangles and circles correspond to the output pulse width measured at the first EAM and the second EAM, respectively. The signal wavelength was 1550.7 nm. RF power was set at 16 dBm and

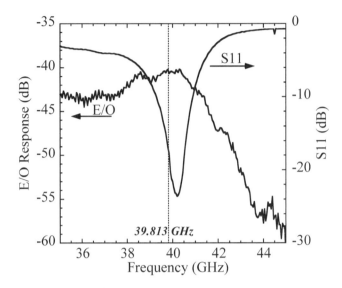

Figure 2.28 Modulation E/O response and return loss characteristics of an SPM-EA modulator [61]. (Reproduced by permission of ©2005 IEICE)

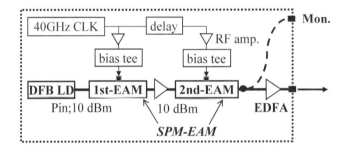

Figure 2.29 Optical pulse source with two-stage EA modulators [61]. (Reproduced by permission of ©2005 IEICE)

the applied bias was kept the same for each EAM. The average light power launched to both EAMs was set at 10 dBm. The pulse width could be reduced by about 30 % after gating twice, and a signal pulse width of less than 3.2 ps was obtained below the bias of −3 V. Note that a 30 % reduction in pulse width is consistent with the prediction from Equation (2.11). Another important advantage of cascaded EAM structure is in enhancing the pulse extinction ratio. The broken line in Figure 2.30 shows the simulated extinction ratio at the output of the second EA modulator, taking the static bias dependence of absorption into account exactly. The extinction ratio was expected to be more than 30 dB at a bias voltage greater than −3.2 V. Figure 2.31 represents the auto-correlation trace (a) and the optical spectrum (b) of the output pulse train. The bias voltage was set at −3 V, and in this case, the pulse width achieved was 3.1 ps with an extinction ratio of 35 dB. The pulse shape was very close to Gaussian and the time–bandwidth product was estimated at 0.5. The slightly greater time bandwidth product compared with

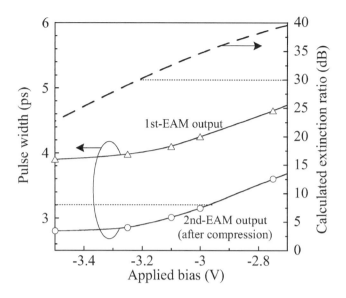

Figure 2.30 Pulse width variation (solid line) and expected extinction ratio (broken line), as a function of applied bias [61]. (Reproduced by permission of ©2005 IEICE)

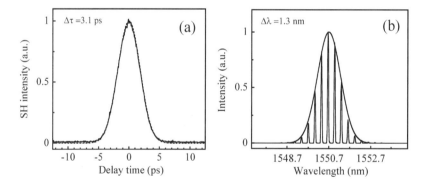

Figure 2.31 (a) Auto-correlation trace and (b) optical spectrum, taken at the output of 2nd-EAM

Fourier transform limit of 0.441 is mainly due to the existence of chromatic dispersion in the two EDFAs, and the chirp characteristics of the EAMs. Further reduction of pulse width, maintaining a good extinction ratio, is easily achieved by increasing the swing voltage V_{pp}.

Due to the simple principle of EAM-based optical pulse generation, the pulse train should reproduce the timing jitter of the modulation signal. As the 40-GHz modulation signal fed by a synthesized signal generator holds an extremely low timing jitter of about 100 fs, the optical pulse train should also should maintain a sufficiently low timing jitter that meets with the criteria described in Section 2.1.

The EA modulator-based pulse source is also applicable in the wavelength range from 1540 nm to 1560 nm, by changing the driving conditions slightly. Figure 2.32 illustrates

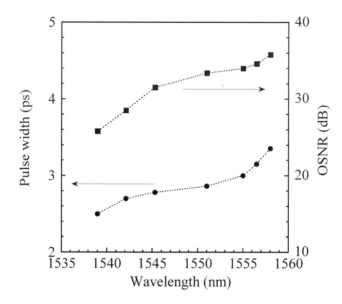

Figure 2.32 Wavelength dependence of output pulse width and OSNR (RSB = 1 nm).

the wavelength dependence of output pulse width and optical signal-to-noise ratio (OSNR). Depending on the nature of the EA modulator, optical loss increases in the shorter wavelength region. Although it led to increase an ASE (amplified spontaneous emission) noise in the signal, an OSNR greater than 28 dB (resolution band width: 1 nm), which was high enough to achieve a Q-factor of more than 30 dB in the 160 Gb/s-OTDM signal, was maintained even at a wavelength of 1540 nm. Such wide-band applicability could be a great advantage for the OTDM/WDM hybrid system.

2.3.3 Optical Time Division Multiplexer Based on EAMs

An optical multiplexer, which converts optical pulse trains to high-speed OTDM signals, performs data modulation for each multiplexed channel individually, thus requiring multiple optical modulators. It is important that these modulators are efficiently installed within a compact package, and for this reason, an optical multiplexer based on free-space integration technology would be a good candidate. Figure 2.33 shows (a) a schematic diagram and (b) a general view of the 4×40 Gb/s (39.82312 Gb/s) optical multiplexer [10, 61, 65]. The optical modulators adopted here are lens-coupled modular EAMs, which are small and offer high-speed modulation with a bandwidth of 35 GHz. The four EA modulators are individually encapsulated in hermetically sealed packages to secure stability and reliability in modulation and are placed in four deferent spatial paths composed of half mirrors and total-reflection mirrors. Each of four 40-Gb/s electrical data signals is directly supplied to the respective modulator by connecting a coaxial cable with a V-band connector. As there exists no electromagnetic crosstalk between electrical data signals, this greatly reduces the complexity in a high-speed electrical design. The modulators each convert an input 40-GHz optical pulse-train into 40-Gb/s optical signals. Each of the four 40-Gb/s optical signals is sequentially bit-interleaved to 80-Gb/s and then to

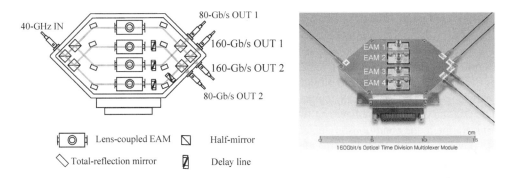

Figure 2.33   (a) Schematic diagram and (b) general view of the 4 × 40-Gb/s optical multiplexer [65]. (Reproduced by permission of ©2007 IEEE)

160-Gb/s, and two sets of 80-Gb/s signal and 160-Gb/s signal are then output. Two 80-Gb/s outputs and a 160-Gb/s output (80-Gb/s OUT 1, 2, 160-Gb/s OUT 2) are used as reference signals for the alignment of modulation timing and for a carrier-phase monitoring. The insertion loss of the optical multiplexer is 15 dB and the loss deviation between multiplexed channels is less than 0.5 dB. The delay-time accuracy of bit interleaving was estimated to be better than 10 fs (6.28 ± 0.01 ps).

The developed optical multiplexer is unique in that the optical carrier phases of all multiplexed channels are controlled separately. The optical carrier-phase difference between adjacent bits is an important factor in fiber transmission, since nonlinear pulse-to-pulse interaction depends on the phase correlation. However, the carrier-phase difference, particularly in a fiber-based optical multiplexer, is usually indefinite and unstable, since the multiplexed channels experience different effective optical lengths, which are much longer than the signal wavelength and are sensitive to variations in environmental conditions. Therefore, the transmission performance could become unstable. In the optical multiplexer illustrated in Figure 2.33, undesirable variation in the effective optical length for each multiplexed channel is small enough, thanks to the free-space design. Thus, the carrier-phase relationship between the multiplexed channels is kept at a stable constant. The stability of the carrier-phase was estimated to be 0.08 radians/°C. A more beneficial feature of the optical multiplexer is that the relative carrier phases of multiplexed channels can be optionally controlled with changing operation temperatures of the EA modulators. As the temperature change brings about a deviation in effective refractive index in an EA waveguide of $5 \times 10^{-4}[1/°C]$, a phase shift of π radians should be achieved with a temperature change of 8 °C under the assumption of a 200 μm-long EA waveguide. Figure 2.34 represents the output power variation of 80-Gb/s OUT 2 as a function of the operating temperatures of EAM 2 and EAM 4. The output power was measured by inputting CW light instead of a pulsed signal. A signal wavelength was set at 1548 nm, and the bias voltage applied to each EA modulator was set at 0 V. The output power varied sinusoidally depending on the temperature of EAM 4, and the condition minimizing interferometer output moved with increase in the temperature of EAM 2. This indicated that the relative carrier phase between multiplexed signals can be freely adjusted by the temperature change. Insertion loss change of EA modulators is smaller than 1 dB under operational temperatures ranging from 15 °C to 35 °C. Thanks to this feature, the optical multiplexer enables generation of phase-coded signals such as CS-RZ signals, even at a bit rate of 160 Gb/s. The CS-RZ modulation format [66] is well-known to be tolerable to nonlinear pulse-to-pulse inter-actions and could contribute to improvements of 160-Gb/s transmission performance [10,61]. In order to generate 160-Gb/s CS-RZ signals, a continuous phase state is imposed on each of two 80-Gb/s signals prior to bit interleaving. The continuous-phase, 80-Gb/s signal can be easily obtained by adjusting thetemperatures driving the two EAMs, as shown in Figure 2.34. The relative carrier phase between two 80-Gb/s signals can be aligned by adding an offset temperature to the conditions giving continuous carrier-phase. Figure 2.35 illustrates CW out-put power variation when the offset temperature was added to the operation temperatures of EAM 2 and EAM 4. The temperatures driving EAM 2 and EAM 4 were initially tuned at 25 °C and 28.5 °C, and were then given the offset temperature while maintaining the continuous-phase condition. Here, the temperatures of EAM 1 and EAM 3 were fixed at 21.5 °C and 21 °C, respectively. As the relative carrier phase between two multiplexed beams should vary depending on the offset-temperature; ΔT, the output power at 160-Gb/s ports changes sinus-oidally, as can be seen in Figure 2.35. In accordance with the nature of the interferometer, the output powers launched from two 160-Gb/s output ports are complementary to each other, so that a continuous-phase signal and a π-radians out-of-phase signal should be launched simultaneously when a pulsed-signal is input. For precise control of the carrier-phase, it is important to monitor the carrier-phase relationship with great accuracy. The optical phase state of each output can be estimated by a free-space, 1-bit-delay interferometer, which is widely used as optical demodulator for DPSK signals. Since the output power of the interferometer strongly depends on the carrier-phase difference between adjacent bits, the phase information is transformed into a function of optical power. By using the phase monitoring method, we have also developed an adaptive phase controller, which has been described elsewhere [67].

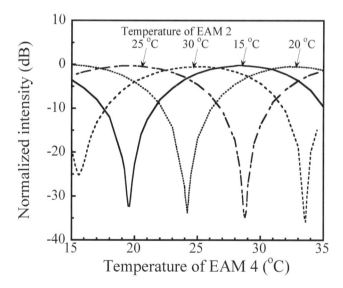

Figure 2.34 Output power variation of 80-Gb/s OUT 2 as a function of the operation temperatures of EAM 2 and EAM 4 [65]. (Reproduced by permission of ©2007 IEEE)

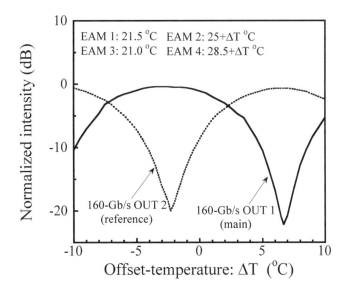

Figure 2.35 CW output power variation at 160-Gb/s ports as a function of the offset temperature [65]. (Reproduced by permission of ©2007 IEEE)

2.3.4 160-Gb/s Optical Signal Generation

Figure 2.36 shows the block diagram of a 160-Gb/s OTDM transmitter, consisting of an optical pulse source and an optical multiplexer. The optical pulse source generates a 40-GHz pulse train, and is launched to the 160-Gb/s optical multiplexer. At the multiplexer, the pulse stream is converted into four 40-Gb/s data streams and then bit-interleaved to a 160-Gb/s signal. The two EAMs are driven by the 40 (39.81312)-GHz clock, with adjusted timing, and convert the CW beam into a Gaussian pulse train. The bias voltage applied to each EAM was -3 V, and an RF power of modulation signal was 20 dBm. The average light powers launched to both EAMs were set at 10 dBm. The pulse width was about 2.8 ps and the extinction ratio was expected to be about 35 dB. A time–bandwidth product of 0.46, which was very close to the Fourier transform limit of a Gaussian pulse, was achieved by carefully compensating residual chirp with short pieces of optical fiber. At the optical multiplexer, a 40-GHz pulse train is converted to four streams of 40-Gb/s optical data, and then bit-interleaved to 160 (159.2525)-Gb/s OTDM signals. As described in the previous section, the optical multiplexer enables the generation of phase-coded OTDM signals. Optical spectra and optical sampling traces shown in Figure 2.37 demonstrate simultaneous generation of CS-RZ signals (160-Gb/s OUT 1) and continuous-phase RZ signals (160-Gb/s OUT 2). The operation temperatures of four EAMs were adjusted to 20 °C, 31 °C, 22 °C, and 35 °C, respectively. The 40-Gb/s ETDM signals with the pseudo-random-bit-sequence (PRBS) of $2^{15} - 1$ were used to modulate the four EAMs. The electric power of the 40-Gb/s data signal was 14 dBm, and the bias voltage applied to EAM was typically -1.5 V, which was slightly different depending on the operation temperature. The CS-RZ coding is dealt with in such a way that the optical pulses acquire the phase alternating between 0 and π in every time slot, so that the carrier component in the CS-RZ signal

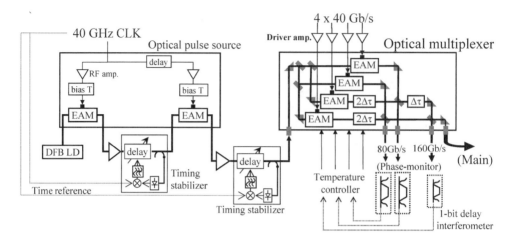

Figure 2.36 Configuration of a 160-Gb/s OTDM transmitter [65]. (Reproduced by permission of ©2007 IEEE)

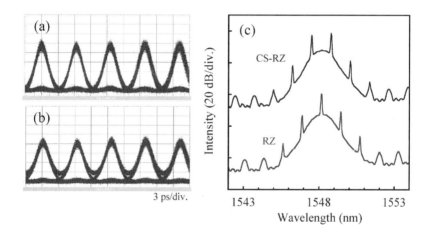

Figure 2.37 Optical sampling traces (a), (b) and optical spectra (c) of a 160-Gb/s CS-RZ signal and 160-Gb/s RZ signal [65]. (Reproduced by permission of ©2007 IEEE)

spectrum is to be canceled out. The optical spectrum exhibits the spectral shapes peculiar to their respective modulation formats, and there is no clear evidence of modulation-data correlation, thanks to the independent encoding. Although the slight nonuniformity of pulse-peak power was observed, the eye diagrams showed clear eye opening. Recent theoretical and experimental work has predicted that effective suppression of intrachannel four-wave-mixing, which is known to be one of the strong limiting factors in fiber-optic transmission [68], is achieved using new modulation formats [69–71]. The modulation scheme is characterized by an alternating phase between 0 and $\pi/2$, i.e. '0 $\pi/2$ 0 $\pi/2$' ($\pi/2$-APRZ), or pair-wise alternating phase between 0 and π, i.e. '0 0 π π' (PAP-CSRZ), and

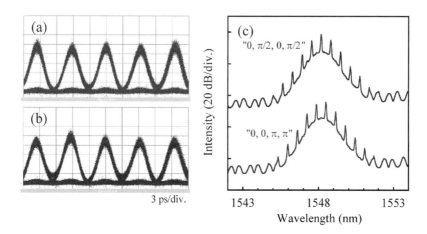

Figure 2.38 Optical sampling traces (a), (b) and optical spectra (c) of 160-Gb/s signal with a phase code of '0 $\pi/2$ 0 $\pi/2$' and '0 0 π π' [65]. (Reproduced by permission of ©2007 IEEE)

is also realizable utilizing the carrier-phase tunability of the developed optical multiplexer. Figure 2.38 shows optical sampling waveforms and optical spectra of such phase-coded signals. The $\pi/2$-APRZ signal having a phase code of '0 $\pi/2$ 0 $\pi/2$' corresponds to a transient state transforming a continuous-phase signal into a π-radian out-of-phase signal. So, both carrier component and sideband components 80 GHz away are observed in the optical spectrum. The pulse tail overlapping in the eye diagram is larger than CS-RZ, but smaller than continuous-phase RZ. In the PAP-CSRZ signal phase-coded with '0 0 π π', the relative phase alternates between 0 and π with the period of 80 GHz, so that the spectrum is equivalent to an 80-Gb/s CS-RZ. As shown in the eye diagram, large pulse overlapping, like continuous-phase RZ, and no overlapping, like CS-RZ, alternates. The 160-Gb/s signal with $\pi/2$-APRZ can be generated only by setting the relative phase difference between two continuous-phase 80-Gb/s signals at 0 or the multiple of 2π, prior to bit-interleaving to 160-Gb/s. Regarding the case of PAP-CSRZ, the 80-Gb/s signals to be bit-interleaved must be CS-RZ with a relative phase difference of 0 or a multiple of 2π. In any case, two 160-Gb/s signals with the same modulation format should be launched from the multiplexer, unlike the case of CS-RZ generation.

2.3.5 Detection of a 160-Gb/s OTDM Signal

Back-to-back performance of a 160-Gb/s OTDM signal was evaluated using a 160-Gb/s receiver consisting of an EA modulator-based optical demultiplexer and a clock recovery circuit. The configuration of the receiver is shown in Figure 2.39. An incoming 160-Gb/s signal is optically demultiplexed to 40 Gb/s by the EA modulator driven by a 40-GHz clock, and then is sent to a 40-Gb/s receiver. Strictly speaking, four EA modulators are required for 1:4 optical demultiplexing as well as the multiplexer. Also, in this case, the authentic demultiplexing module will be realizable by employing free-space structure similar to the optical multiplexer. Figure 2.40 illustrates the optical sampling traces of an optically demultiplexed 40-Gb/s signal. The gate width of the demultiplexer was measured at 4 ps by inputting CW

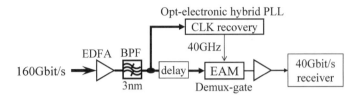

Figure 2.39 160-Gb/s OTDM receiver employing a polarization insensitive EAM-based optical demultiplexer [65]. (Reproduced by permission of ©2007 IEEE)

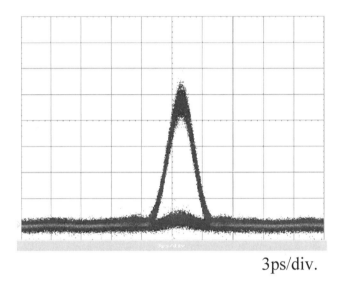

3ps/div.

Figure 2.40 Optical sampling traces of an optically demultiplexed 40-Gb/s signal [65]. (Reproduced by permission of ©2007 IEEE)

light instead of a 160-Gb/s signal. The sampling trace shows that the 40-Gb/s signal was clearly separated from the 160-Gb/s signal. The suppression ratio for neighboring channels was estimated to be more than 15 dB. Since the polarization state of the incoming signal is not always constant, the polarization insensitivity is an essential requirement for stable operation of an optical demultiplexer. The polarization dependent loss (PDL) of the EA modulator is designed to be less than 0.5 dB by applying tensile strain to an InGaAsP-based MQW electroabsorption layer, so that polarization-insensitive demultiplexing is secured.

The 40-GHz clock signal is recovered with an optical–electronic, hybrid phase, locked-loop (H-PLL) shown in Figure 2.41 [61, 65, 72]. Here, a polarization-insensitive EA modulator is also installed at the front-end of the PLL circuit and is driven by a locally generated clock signal, which is derived from a voltage controlled oscillator (VCO). As a small frequency offset, Δf (250 MHz) is added to the clock signal, the output optical signal from the EA modulator includes the beat frequency component of $4\Delta f$ (1 GHz). The optical output beating at $4\Delta f$ is converted to an electrical signal using an O/E converter (3-dB bandwidth = 1.7 GHz), and a band pass filter (BPF, center frequency = 1 GHz). Then the phase deviation between the prescaled signal and the reference signal ($4\Delta f$) coming from a local oscillator (LO) is detected at a phase comparator to facilitate PLL operation. The root mean square (RMS) timing jitter

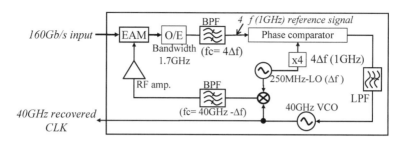

Figure 2.41 Opt-electronic hybrid PLL circuit using polarization insensitive EAM as a prescaler [65]. (Reproduced by permission of ©2007 IEEE)

of the recovered 40-GHz clock was approximately 60 fs. The capture range and the locking range of PLL operation were measured at 1.2 MHz and 12 MHz, respectively. The robustness of the PLL operation was evaluated by checking for 160-Gb/s signals distorted by chromatic dispersion or PMD, and it was confirmed that the locking characteristic was still stable within a range of chromatic dispersion between −5 ps/nm and 5 ps/nm, and within a range of DGD (differential group delay) by PMD between 0 ps and 3 ps.

Figure 2.42 illustrates the back-to-back performance of the developed equipment. The bit error rate (BER) was measured for each of the four 40-Gb/s demultiplexed signals, thus varying the decision level in bit-error evaluation. At the transmitter, the PRBS of four 40-Gb/s ETDM signals was set at $2^{15}-1$. A 160-Gb/s signal was tuned to a CS-RZ signal. Figure 2.42 shows the

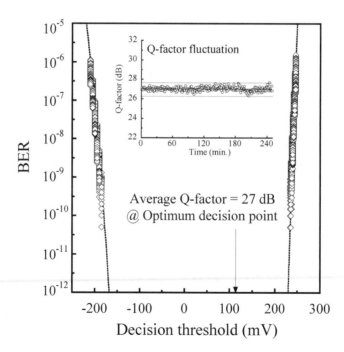

Figure 2.42 Back-to-back BER performance evaluated for 4 hours. Average Q-factor was estimated at 27 dB. Inset illustrates the time-dependent Q-factor fluctuation [65]. (Reproduced by permission of ©2007 IEEE)

results for a 40-Gb/s tributary that exhibited the worst BER performance. The measurement was performed continuously over 4 hours, and the Q-factor [73], estimated from the BER measurement, was 27 dB on average. This value corresponds to a BER value of 10^{-100} or less at the optimum decision point, attesting to excellent back-to-back performance. The Q-factor fluctuation during 4-hour continuous measurement was less than 1.3 dB, which revealed an extremely stable performance for the EAM-based OTDM system, an essential factor in practical applications.

The same evaluation of the signal wavelengths of 1540 nm and 1560 nm also revealed equivalent back-to-back performances. Table 2.1 summarizes Q-factors and optical signal-to-noise ratios (OSNR), measured at wavelengths of 1540 nm, 1548 nm, and 1560 nm. The OSNR was measured at the transmitter end. It was confirmed that Q-factors greater than 26 dB were obtained over a wavelength range of 20 nm. Slightly worse Q-factors for shorter wavelengths originate in OSNR degradation due to larger insertion loss in the EAM. It is thought that the result proves the wide-band applicability of the EA modulator based OTDM system quantitatively. The back-to-back performances of the other three modulation formats described above are also summarized in Table 2.2. The Q-factors of each modulation format exhibits very close values, suggesting that the developed prototype equipment is useful for transmitting ultra high-speed optical signals with various modulation formats over a wide wavelength range.

Table 2.1 Wavelength dependence of back-to-back performance on CS-RZ signal [65] (Reproduced by permission of ©2007 IEEE)

Wavelength	1540 nm	1548 nm	1560 nm
OSNR at Res. = 1 nm	27 dB	30 dB	32 dB
Q-factor at worst tributary	26.1 dB	27.1 dB	28.3 dB

Table 2.2 Back-to-back performance of different modulation formats (signal wavelength = 1548 nm) [65] (Reproduced by permission of ©2007 IEEE)

Modulation format	RZ	$\pi/2$-APRZ	PAP-CSRZ
Q-factor at worst tributary	26.5 dB	27 dB	26.7 dB

2.3.6 Transmission Issues

Recent ultra-high-speed transmission experiments using installed optical fiber cables have clearly demonstrated the necessity of applying adaptive compensation of polarization mode dispersion (PMD) [13,74–76]. Unlike in relatively stable laboratory environments, for installed optical fibers in practical use, PMD and chromatic dispersion varies according to external factors such as weather conditions [74]. For practical applications, the desired transmission characteristics must be maintained under circumstances in which the optical fiber characteristics are subject to continuous change. In such situations, the stability and high performance of transmission equipment becomes more and more important, especially with increasing bit rates. The EAM-based OTDM technology described here offers a robust operation for 160-Gb/s signal generation, detection, and a switching capability in modulation format. In order to reveal how it contributes to stable transmission in a deployed system, we performed a field transmission experiment using an optical test bed in the Japan Gigabit Network II (JGNII). It successfully demonstrated a 635-km long-distance transmission employing a robust 160-Gb/s CS-RZ OTDM signal. The trial results are briefly reviewed in Chapter 7. Also the detail of field trial is described in Ref. [65].

2.4 Summary

Ultra-high-speed signal generation is the most fundamental technology in the realization of next-generation, ultra-high-capacity photonic networks. In this chapter, we have described two types of ultra-high-speed signal sources based on monolithic types of MLLD and EA modulators.

An active MLLD has excellent features such as a wide locking range, although the repetition frequency suffers from electrical limitation. It is a promising candidate for optical signal sources for 40-Gb/s RZ transmission systems and 160-Gb/s OTDM systems. A passive MLLD has the possibility of generating higher repetition frequency and narrower pulse width because of release from electrical limitation. At present, a repetition frequency of over 1 THz [40] and the pulse width of 0.8 ps without pulse compression [54] have been reported. The drawback of a passive MLLD was its not being synchronized with external electrical signals and comparably large timing jitter. However, this has been solved by developing the stabilization techniques of hybrid mode locking and optical synchronous mode locking. It is possible to generate ultrafast and stable optical pulses at over 100 GHz by subharmonic synchronous mode locking. Passive MLLDs are expected to be key devices for future, ultrafast, all-optical communications systems.

EAM-based optical pulse sources have the ability stably to generate optical short pulse trains using very simple operation principles. Thanks to cascaded modulation with two EAMs, the generated pulse shows a sufficiently high extinction ratio with low timing jitter. Such characteristics are very suitable for 160-Gb/s OTDM signal transmission. EAMs are also introduced to 160 Gb/s (4×40 Gb/s) optical time division multiplexers utilizing free-space integration. The 160-Gb/s optical multiplexer supports individual data encoding for each of all multiplexed signals and have established a technique for generating stable, high-quality, 160-Gb/s OTDM signals while maintaining both operability and practicability. CS-RZ is an example of a superior modulation format suitable for ultra-high-speed transmission. The EAM-based optical multiplexer has enabled the introduction of such formats into 160-Gb/s signals, taking advantage of the stable carrier phase, a characteristic of space-integrated optical multiplexers. These are regarded as core technologies for generating ultra-high-speed OTDM exceeding the processing speeds of electronic devices.

References

[1] K. Murata, K. Sano, H. Kitabayashi, S. Sugitani, H. Sugahara, and T. Enoki, '100 Gb/s multiplexing and demultiplexing IC operations in InP HEMT technology', *IEEE J. Solid-State Circuits*, **39**, 207–213 (2004).
[2] P. J. Winzer, G. Raybon, and M. Duelk, '107-Gb/s optical ETDM transmitter for 100G Ethernet transport', *Proceedings 31st European Conference on Optical Communications,* (ECOC2006), Glasgow, post-deadline paper 4.1.1 (2005).

[3] Y. Suzuki, Z. Yamazaki, Y. Amamiya, S. Wada, H. Uchida, C. Kurioka, S. Tanaka, and H. Hida, '120 Gb/s multiplexing and 110 Gb/s demultiplexing ICs', *IEEE J. Solid-State Circuits*, **39**, 2397–2402 (2004).

[4] M. Nakazawa, T. Yamamoto, and K. R. Tamura, '1.28 Tbit/s–70 km OTDM transmission using third- and fourth-order simultaneous dispersion compensation with a phase modulator', *Proceedings 26th European Conference on Optical Communications* (ECOC2000), Munich, Germany, post-deadline paper PD2.6 (2000).

[5] H. G. Weber, S. Ferber, M. Kroh, C. Schmidt-Langhorst, R. Ludwig , V. Marembert, C. Boerner , F. Futami, S. Watanabe, and C. Schubert 'Single channel 1.28 Tbit/s and 2.56 Tbit/s DQPSK transmission', *Proceedings 31st European Conference on Optical Communications* (ECOC2006), Glasgow, post-deadline paper 4.1.2 (2005).

[6] S. Kawanishi, H. Takara, K. Uchiyama, I. Shake, and K. Mori, '3Tbit/s (160 Gbit/s × 19ch) OTDM-WDM transmission experiment,' *Technical Digest, Optical Fiber Communications Conference 1999* (OFC1999), San Diego, USA, post-deadline paper PD1 (1999).

[7] A. Suzuki, X. Wang, T. Hasegawa, Y. Ogawa, S. Arahira, K. Tajima, and S. Nakamura, '8 × 160 Gb/s (1.28 Tb/s) DWDM/OTDM unrepeated transmission over 140 km standard fiber by semiconductor-based devices', *Proceedings 29th European Conference on Optical Communications* (ECOC2003), Rimini, Italy, paper Mo3.6.1, pp. 44–47 (2003).

[8] E. Lach, K. Schuh, M. Schmidt, B. Junginger, G. Charlet, P. Pecci, and G. Veith, '7 × 170 Gbit/s (160 Gbit/s + FEC overhead) DWDM transmission with 0.53 bit/s/Hz spectral efficiency over long haul distance of standard SMF', *Proceedings 29th European Conference on Optical Communications* (ECOC2003), Rimini, Italy, post-deadline paper Th4.3.5, pp. 68–69 (2003).

[9] L. Möller, Y. Su, X. Liu, J. Leuthold, and C. Xie, 'Generation of 160 Gb/s carrier-suppressed return-to-zero signals', *Proceedings 29th European Conference on Optical Communications* (ECOC2003), Rimini, Italy, paper Mo3.6.3, pp. 50–51 (2003).

[10] H. Murai, M. Kagawa, H. Tsuji, and K. Fujii, 'Single channel 160 Gbit/s carrier-suppressed RZ transmission over 640 km with EA modulator based OTDM module', *Proceedings 29th European Conference on Optical Communications* (ECOC2003), Rimini, Italy, paper Mo3.6.4, pp. 52–54 (2003).

[11] S. Feber, R. Ludwig, C. Boerner, A. Wietfeld, B. Schmauss, J. Berger, C. Schubert, G. Unterboersch, and H. G. Weber, 'Comparison of DPSK and OOK modulation format in a 160 Gb/s transmission system', *Proceedings 29th European Conference on Optical Communications* (ECOC2003), Rimini, Italy, paper Th2.6.2, pp. 1004–1005 (2003).

[12] T. Inui, K. Mori, T. Ohara, H. Takara, T. Komukai, and T. Morioka, '160 Gbit/s adaptive dispersion equalizer using a chirp monitor with a balanced dispersion configuration', *Proceedings 29th European Conference on Optical Communications* (ECOC2003), Rimini, Italy, paper Tu3.6.3, pp. 266–267 (2003).

[13] S. Kieckbusch, S. Ferber, H. Rosenfeldt, R. Ludwig, C. Boerner, A. Ehrhardt, E. Brinkmeyer, and H.-G. Weber, 'Automatic PMD compensator in a 160-Gb/s OTDM transmission over deployed fiber using RZ-DPSK modulation format', *J. Lightwave Technol.*, **23**, 165–171, (2005).

[14] S. Watanabe, F. Futami, R. Okabe, and Y. Takita, '160 Gbit/s optical 3R-regenerator in a fiber transmission experiment', *Technical Digest, Optical Fiber Communications Conference 2003* (OFC2003), Atlanta, USA, post-deadline paper PD16 (2003).

[15] J. P. Gordon and H. A. Haus, 'Random walk of coherently amplified solitons in optical fiber transmission', *Optical Letters,* **11**, 665–667 (1986).

[16] E. Iannone, F. Matera, A. Mecozzi, and M. Settembre, *Nonlinear Optical Communication Networks*, John Wiley & Sons, Inc., New York, 1998.

[17] M. Jinno, 'Effect of timing jitter on an optically controlled picosecond optical switch', *Optical Letters*, **18**, 1409–1411 (1993).

[18] R. Ludwig, S. Dies, A Ehrhardt, L. Kuller, W. Pieper, and H. G. Weber, 'A tunable femtosecond modelocked semiconductor laser for applications in OTDM systems', *IEICE Trans. Electron.*, **E81-C**, no.2 pp. 140–145 (1998).

[19] K. Sato, A. Hirano, and H. Ishii, 'Chirp-compensated 40-GHz mode-locked lasers integrated with electroabsorption modulators and chirped gratings', *IEEE J. Selected Topics Quantum Electron.*, **5**, 590–595 (1999).

[20] Y. Hashimoto, H. Yamada, R. Kuribayashi, and H. Yokoyama, '40-GHz tunable optical pulse generation from a highly stable external-cavity mode-locked semiconductor laser module', *Technical Digest, Optical Fiber Communications Conference 2002* (OFC2002), Anaheim, USA, paper WV5 (2002).

[21] S. Arahira and Y. Ogawa, '40GHz actively mode-locked distributed Bragg reflector laser diode module with an impedance-matching circuit for efficient RF signal injection', *Jpn. J. Appl. Physics*, **43**, 1960 (2004).

[22] J.-L. Augé, M. Cavallari, M. Jones, P. Kean, D. Watley, and A. Hadjifotiou, 'Single channel 160 GB/s OTDM propagation over 480 km of standard fiber using a 40 GHz semiconductor mode-locked laser pulse source', *Technical Digest, Optical Fiber Communications Conference 2002* (OFC2002), Anaheim, USA, paper TuA3 (2002).

[23] H. Murai, M. Kagawa, H. Tsuji, K. Fujii, Y. Hashimoto, and H. Yokoyama, 'Single channel 160 Gbit/s (40 Gbit/s × 4) 300 km transmission using EA modulator based-OTDM module and 40 GHz external-cavity mode-locked LD', *Proceedings 28th European Conference on Optical Communications* (ECOC2002), Copenhagen, Denmark, paper 2.1.4 (2002).

[24] A. Suzuki, X. Wang, T. Hasegawa, Y. Ogawa, S. Arahira, K. Tajima, and S. Nakamura, '8 × 160 Gb/s (1.28 Tb/s) DWDM/OTDM unrepeated transmission over 140 km standard fiber by semiconductor-based devices', *Proceedings 29th European Conference on Optical Communications* (ECOC2003), Rimini, Italy, paper Mo3.6.1, pp. 44–47 (2003).

[25] M. Suzuki, H. Tanaka, N. Edagawa, K. Utaka, and Y. Matsushima, 'Transform-limited 14 ps optical pulse generation with 15 GHz repetition rate by InGaAsP electroabsorption modulator', *Electron. Lett.*, **28**, 1007–1008 (1992).

[26] E. Lach, K. Schuh, and M Schmidt, 'Application of electroabsorption modulators for high-speed transmission systems', *Journal of Optical and Fiber Communications Reports*, **2**(2), 140–170 (2005).

[27] A. R. Pratt, H. Murai, H. T. Yamada, and Y. Ozeki, '8 × 80 Gbit/s electrical TDM-transmission over 640 km of large effective area non-zero dispersion shifted fiber', *Proceedings 26th European Conference on Optical Communications* (ECOC2000), Munich, Germany, paper 10.1.6, pp. 29–30 (2000).

[28] D. A. B. Miller, D. S. Chemla, T. C. Damen, A. C. Gossard, W. Wiegmann, T. H. Wood, and C. A. Burrus 'Band-edge electroabsorption in quantum well structures: The quantum-confined stark effect', *Phys. Rev. Lett.*, **53**, 2173–2176 (1984).

[29] G. Raybon, P. J. Winzer, and C. R. Doerr, '10 × 107-Gbit/s electronically multiplexed and optically equalized NRZ transmission over 400 km', *Technical Digest, Optical Fiber Communications Conference 2006* (OFC2006), Anaheim, USA, post-deadline paper PD32 (2006).

[30] M. Doi, S. Taniguchi, M. Seino, G. Ishikawa, H. Ooi, and H. Nishimoto, '40 Gb/s integrated OTDM Ti:LiNbO3 modulator,' in *Proceedings 1996 Photonics Switching* (PS96), Sendai, Japan, PthB1, pp. 172–173 (1996).

[31] A. Hirano, M. Asobe, K. Sato, Y. Miyamoto, K. Yonenaga, H. Miyazawa, M. Abe, H. Takara, and I. Shake 'Dispersion tolerant 80-Gbit/s carrier-suppressed return-to-zero (CS-RZ) format generated by using phase- and duty-controlled optical time division multiplexing (OTDM) technique', *IEICE Trans. Communications*, **E85-B**, No.2, pp. 431–437 (2002).

[32] T. Ohara, H. Takara, I. Shake, K. Mori, S. Kawanishi, S. Mino, T. Yamada, M. Ishii, T. Kitoh, T. Kitagawa, K. R. Parameswaran, and M. M. Fejer, '160-Gb/s optical-time-division multiplexing with PPLN hybrid integrated planner lightwave circuit', *IEEE Photon. Technol. Lett.*, **15**, 302–304 (2003).

[33] J. E. Bowers, P. A. Morton, A. Mar, and S. W. Corzine, 'Actively mode-locked semiconductor lasers,' *IEEE J. Quantum Electron.*, **25**, 1426–1439 (1989).

[34] H. Yokoyama, 'Highly reliable mode-locked semiconductor lasers,' *IEICE Trans. Electron.*, **E85-C**, No.1, pp. 27–36 (2002).

[35] K. Sato, K. Wakita, I. Kotaka, Y. Kondo, and M. Yamamoto, 'Monolithic strained-InGaAsP multiple-quantum-well lasers with integrated electroabsorption modulators for active mode locking,' *Appl. Phys. Lett.*, **65**, 1–3 (1994).

[36] Y. Katoh, S. Arahira, and Y. Ogawa, '40 GHz actively mode-locked DBR laser diodes with a wide (800 MHz) locking range,' OFC2001, WC5-1, Anaheim, CA (2001).

[37] R. S. Tucker, U. Koren, G. Raybon, C. A. Burrus, B. I. Miller, T. L. Koch, and G. Eisenstein, '40 GHz active mode-locking in a 1.5 mm monolithic extended-cavity laser,' *Electron. Lett.*, **25**, 621–622 (1989).

[38] H. A. Haus, 'Modelocking of semiconductor laser diodes,' *Jpn. J. Appl. Phys.*, **20**, 1007–1020 (1981).

[39] S. Arahira, Y. Matsui, T. Kunii, S. Oshiba, and Y. Ogawa, 'Transform-limited optical short-pulse generation at high repetition rate over 40 GHz from a monolithic passive mode-locked DBR laser diode,' *IEEE Photon. Technol. Lett.*, **5**, 1362–1365 (1993).

[40] Y. K. Chen and M. C. Wu, 'Monolithic colliding-pulse mode-locked quantum-well lasers,' *IEEE J. Quantum Electron.*, **28**, 2176–2185 (1992).

[41] S. Arahira, S. Oshiba, Y. Matsui, T. Kunii, and Y. Ogawa, 'Terahertz-rate optical pulse generation from a passively mode-locked semiconductor laser diode,' *Optical Lett.*, **19**(11), 834–836 (1994).

[42] T. Shimizu, S. Wang, and H. Yokoyama, 'Asymmetric colliding-pulse mode-locking in InGaAsP semiconductor lasers,' *Optical Rev.*, **2**, 401–403 (1995).

[43] H. A. Haus, 'Theory of mode locking with a slow saturable absorber,' *IEEE J. Quantum Electron.*, **11**, 736–746 (1975).

[44] S. Arahira and Y. Ogawa, 'Passive and hybrid modelockings in a multi-electrode DBR laser with two gain sections,' *Electron. Lett.*, **31**, 808–809 (1995).

[45] D. von der Linde, 'Characterization of the noise in continuously operating mode-locked lasers,' *Appl. Phys.*, **B39**, 201–217 (1986).

[46] D. J. Derickson, P. A. Morton, J. E. Bowers, and R. L. Thornton, 'Comparison of timing jitter in external and monolithic cavity mode-locked semiconductor lasers,' *Appl. Phys. Lett.*, **59**, **26**, 3372–3374 (1991).

[47] H. Yokoyama, T. Shimizu, T. Ono, and Y. Yano, 'Synchronous injection locking operation of monolithic mode-locked diode lasers,' *Optical Rev.*, **2**, 85–88 (1995).

[48] S. Arahira and Y. Ogawa, 'Synchronous mode-locking in passively mode-locked semiconductor laser diodes using optical short pulses repeated at subharmonics of the cavity round-trip frequency,' *IEEE Photon. Technol. Lett.*, **8**, 191–193 (1996).

[49] X. Wang, H. Yokoyama, and T. Shimizu, 'Synchronized harmonic frequency mode-locking with laser diodes through optical pulse train injection,' *IEEE Photon. Technol. Lett.*, **8**, 617–619 (1996).

[50] A. Nirmalathas, H. F. Liu, Z. Ahmed, D. Novak, and Y. Ogawa, 'Subharmonic synchronous mode-locking of a monolithic semiconductor laser operating at millimeter-wave frequencies,' *IEEE J. Select. Topics Quantum Electron.*, **3**, 261–269 (1997).

[51] S. Arahira, Y. Katoh, D. Kunimatsu, and Y. Ogawa, 'Stabilization and timing jitter reduction of 160 GHz colliding-pulse mode-locked laser diode by subharmonic-frequency optical pulse injection,' *IEICE Trans. Electron.*, **E83-C**, 966–973 (2000).

[52] S. Arahira, S. Sasaki, K. Tachibana, and Y. Ogawa, 'All-optical 160-Gb/s clock extraction with a mode-locked laser diode module,' *IEEE Photon. Technol. Lett.*, **16**, 1558–1560 (2004).

[53] S. Arahira and Y. Ogawa, '160-Gb/s OTDM signal source with 3R function utilizing ultrafast mode-locked laser diodes and modified NOLM,' *IEEE Photon. Technol. Lett.*, **17**, 992–994 (2005).

[54] Y. Kawamura, K. Wakita, Y. Itaya, Y. Yoshikuni, and H. Asahi, 'Monolithic integration of InGaAs/InP DFB lasers and InGaAs/InAlAs MQW optical modulators', *Electron. Lett.*, **22**, 242–243 (1986).

[55] M. Suzuki, Y. Noda, H. Tanaka, S. Akiba, Y. Kushiro, and H. Issiki, 'Monolithic integration of InGaAsP/InP distributed feedback laser and electroabsorption modulator by vapor phase epitaxy', *IEEE J. Lightwave Technol.*, **LT-5**, 1277–1284 (1987).

[56] H. Kawanishi, Y. Yamauchi, N. Mineo, Y. Shibuya, H. Mural, K. Yamada, and H. Wada, 'EAM-integrated DFB laser modules with more than 40-GHz bandwidth', *IEEE Photon. Technol. Lett.*, **13**, 954–956 (2001).

[57] N. Mineo,. 'More than 50 GHz bandwidth electroabsorption modulator module', *Technical Digest, 5th Optoelectronics Communications Conference* (OECC2000), PD2-7 (2000).

[58] S. Kodama, T. Yoshimatsu, H. Ito, '500 Gbit/s optical gate monolithically integrating photodiode and electroabsorption modulator', *Electron. Lett.*, **40**, 555–556 (2004).

[59] F. Koyama and K. Iga, 'Frequency chirping in external modulators', *IEEE J. Lightwave Technol.*, **6**, 87–93 (1988).

[60] N. Mineo, K. Yamada, K. Nakamura, S. Sasaki, and T. Ushikubo, '60-GHz band electroabsorption modulator module', *Technical Digest, Optical Fiber Communications Conference 1998* (OFC1998), San Jose, USA, paper ThH4, pp. 287–288 (1998).

[61] H. Murai, M. Kagawa, H. Tsuji, and K. Fujii, 'EA modulator-based optical multiplexing/demultiplexing techniques for 160 Gbit/s OTDM signal transmission', *IEICE Trans. Electron.* **E88-C**, 309–318 (2005).

[62] V Kaman, Y. J. Chiu, T. Liljeberg, S. Z. Zhang, and J. E. Bowers, 'Integrated tandem traveling-wave electroabsorption modulators for > 100 Gbit/s OTDM applications', *IEEE Photon. Technol. Lett.*, **12**, 1471–1473 (2000).

[63] M. Kagawa, H. Murai, H. Tsuji, K. Fujii, 'Single channel 40 Gbit/s-based 160 Gbit/s OTDM transmission over 180 km of SMF', *Technical Digest, 7th Optoelectronics Communications Conference* (OECC2002), paper 9B1-2, pp. 20–21 (2002).

[64] K. Schuh, M. Schmidt, E. Lach, B. Junginger, A. Klekamp, G. Veith, P. Sillard, '4 × 160 Gbit/s DWDM/OTDM transmission over 3 × 80 km TeraLightTM – reverse TeraLightTM fibre', *Proceedings 28th European Conference on Optical Communications* (ECOC2002), Copenhagen, Denmark, paper 2.1.2 (2002).

[65] H. Murai, M. Kagawa, H. Tsuji, and K. Fujii, 'EA-modulator-based optical time division multiplexing/ demultiplexing techniques for 160-Gb/s optical signal transmission', *IEEE J. Selected Topics in Quantum Electron.*, **13**, 70–78 (2007).

[66] Y. Miyamoto, A. Hirano, K. Yonenaga, A. Sano, H. Toba, K. Murata, and O. Mitomi, '320 Gbit/s (8 × 40 Gbit/s) WDM transmission over 367-km zero-dispersion-flattened line with 120-km repeater spacing using carrier-suppressed return-to-zero pulse format', *Optical Amplifiers and Their Applications 1999* (OAA1999), post-deadline paper PDP4 (1999).

[67] M. Kagawa, H. Murai, H. Tsuji, and K. Fujii, 'Performance comparison of bitwise phase-controlled 160 Gbit/s signal transmission using an OTDM multiplexer with phase-correlation monitor', *Proceedings 30th European Conference on Optical Communications* (ECOC2004), Stockholm, Sweden, paper We4.P.109 (2004).

[68] P. V. Mamyshev and N. A. Mamysheva, 'Pulse-overlapped dispersion-managed data transmission and intra-channel four-wave mixing', *Optical Lett.*, **24**, 1454–1456 (1999).

[69] R. I. Killey, H. J.Thiele, V. Mikhailov, and P. Bayvel. 'Reduction of intrachannel nonlinear distortion in 40-Gb/s-based WDM transmission over standard fiber', *IEEE Photon. Technol. Lett.*, **12**, 1624–1626 (2000).

[70] S. Randel, B. Konrad, Anes Hodžić, and K. Petermann., 'Influence of bitwise phase changes on the performance of 160 Gbit/s transmission systems', *Proceedings 28th European Conference on Optical Communications* (ECOC2002), Copenhagen, Denmark, paper P3.31 (2002).

[71] L. Möller, Y. Su, C. Xie, R. Ryf, X. Liu, X. Wei, and S. Cabot, 'All-optical phase construction of ps-pulses from fiber lasers for coherent signaling at ultra-high data rates ($\geq$ 160 Gb/s)', *Technical Digest, Optical Fiber Communications Conference 2004* (OFC2004), post-deadline paper PDP20 (2004).

[72] H. T. Yamada, H. Murai, A. R. Pratt, Y. Ozeki, 'Scalable 80 Gbit/s OTDM using a modular architecture based on EA modulators', *Proceedings 26th European Conference on Optical Communications* (ECOC2000), Munich, Germany, paper 1.3.5, pp. 47–48 (2000).

[73] N. S. Bergano, F. W. Kerfoot, C. R. Davidson, 'Margin measurements in optical amplifier systems', *IEEE Photon. Technol. Lett*, **5**, 304–306 (1993).

[74] T. Miyazaki, M. Daikoku, I. Morita, T. Otani, Y. Nagao, M. Suzuki, and F. Kubota, 'Stable 160-Gb/s DPSK transmission using a simple PMD compensator on the photonic network test bed of JGN2,' *9th OptoEletoronics and Communications Conference/3rd International Conference on Optical Internet* (OECC/COIN 2004), 14C3-2, pp. 462–463 (2004).

[75] M. Schmidt, M. Witte, F. Buchali, E. Lach, E. Le Rouzic, S. Salaun, S. Vorbeck, and R. Leppla, '8 × 170 Gbit/s DWDM field transmission experiment over 430 km SSMF using adaptive PMD compensation', *Proceedings 30th European Conference on Optical Communications* (ECOC2004), Stockholm, Sweden, PD H4.1.2 (2004).

[76] G. Lehmann, W. Schairer, H. Rohde, E. Sikora, Y. R. Zhou, A. Lord, D. Payne, J. P. Turkiewicz, E. Tangdiongga, G. D. Khoe, and H. de Waardt, '160 Gbit/s OTDM transmission field trial over 550 km of legacy SSMF', *Proceedings 30th European Conference on Optical Communications* (ECOC2004), Stockholm, Sweden, We1.5.2 (2004).

[77] S. Arahira, Y. Katoh, and Y. Ogawa, '20 GHz subpicosecond monolithic modelocked laser diode,' *Electron. Lett.*, **36**,.454–456 (2000).

3

Semiconductor Optical Amplifier Based Ultrafast Signal Processing Devices

Hidemi Tsuchida and Shigeru Nakamura

3.1 Introduction

The semiconductor optical amplifier (SOA) is one of the most frequently used devices for ultrafast signal processing. This is because the device technologies are already well established and it is commercially available. For ultrafast signal processing, the key is how to extract the advantageous features of SOA. In this chapter we briefly review the basic properties of SOA, and then discuss its features as an ultrafast nonlinear medium. We introduce two ways of using SOA for ultrafast signal processing. One is to use only the very fast response component of SOA by using a wavelength filter, and the other is to cancel out the slow response component of SOA using symmetric Mach–Zehnder configuration. For each method, we show the application to the ultrafast all-optical signal processing.

3.2 Fundamentals of SOA

The basic structure of an SOA is the same as that of semiconductor laser diodes, except for an anti-reflection (AR) coating on both facets and sometimes the tapered structure of the waveguide close to the facets. These are to minimize the reflection at both facets. The active layer, which can either be bulk, quantum well, or quantum dots, is embedded as a gain medium

Ultrafast All-Optical Signal Processing Devices   Edited by Hiroshi Ishikawa
© 2008 John Wiley & Sons, Ltd

in the form of an optical waveguide. Optical gain of the SOA with bulk or quantum well active layer can be obtained using the first-order susceptibility obtained in Chapter 1 as:

$$g(\omega) = -\frac{\omega}{cn} \int_{k=0}^{k=\infty} D(k) \left[\text{Im} \left(\chi^{(1)}(k) \right) \right] dk$$

$$= \frac{\omega}{cn\varepsilon_0} \int_{k=0}^{k=\infty} D(k) \frac{\hbar \gamma_{CV} |\mu_{cv}|^2}{\hbar^2 (\omega_{cv}(k) - \omega)^2 + \hbar^2 \gamma_{CV}^2} (f_c(k) - f_v(k)) \, dk$$

(3.1)

where ω is the light wave angular frequency, c is the velocity of light, n is the refractive index and ε_0 is the vacuum dielectric constant, $D(k)$ is the density of state expressed in terms of wave number k, which depends on the active layer geometry, bulk or quantum well. $\omega_{CV}(k)$ is the k-dependent transition frequency between conduction band (C) and valence band (V), γ_{CV} is the dephasing time and f_C, f_V are Fermi–Dirac distribution functions. The dipole moment $|\mu_{CV}|^2$ depends on the polarization TE or TM, and contributing valence bands; heavy hole or light hole [1]. Summation of Equation (3.1) over bands is needed when multiple valence bands contribute to the gain. In the case of quantum dot active layer, (3.1) applies for isotropic spherical dot, however, anisotropic dot requires some modification in the integration. Figure 3.1 shows calculated examples of the optical gain for various injection current densities taking the examples of the quantum dot active layer and bulk InGaAsP active layer [2]. We can see the increase in the optical gain by increasing the injection current. The quantum dot active layer exhibits a discrete gain peak due to its three-dimensional quantization.

The key issues in the development of SOA were how to realize large output saturation power, polarization insensitive gain, and small noise figure. The output power saturation takes place

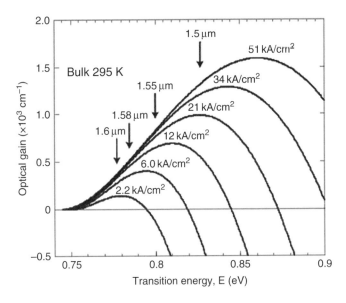

Figure 3.1 Examples of calculated optical gain of SOA. Case of bulk active layer and quantum dot active layer are shown [2]. (Reproduced by permission of © 2001 The Institute of Pure and Applied Physics (IPAP))

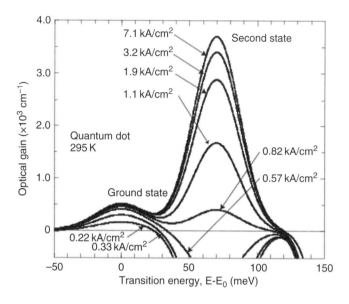

Figure 3.1 (*continued*)

because of the reduced population inversion caused by the intense optical power that consumes the population inversion carriers. If we use the SOA at an output power above the saturation power, we cannot get good eye opening for random pattern modulation, i.e. the pattern effect. Generally the saturation power is given by [3]:

$$P_s = C\hbar\omega\frac{dw}{\Gamma}\frac{1}{g_d\tau}$$ (3.2)

Where C is the fiber-chip coupling efficiency, d is the thickness of active layer, w is the width of active layer, Γ is the optical confinement factor, g_d is differential gain, τ is carrier lifetime. dw/Γ corresponds to the mode cross-section. Efforts have been made to realize large saturation power by the design of the active layer, optical confinement factors and carrier lifetime. As for polarization dependence, a waveguide with a thin active layer gives a small loss to the TE mode when compared with that for the TM mode, resulting in polarization dependence. The rectangular cross section of the waveguide, including the active layer, provides polarization independent waveguide loss. The active layer gain also depends on the polarization. In the bulk active layer the polarization dependence is small because of degenerate heavy-hole and light-hole bands. In the quantum well active layer, the degeneracy is removed and results in polarization-dependent optical gain. The conduction band to light-hole band transition gives gain mainly to the TM mode and that to the heavy-hole gives gain to the TE mode. There are various designs of active layer for minimizing polarization dependence. They include geometry and strain of active layer and the waveguide structure. As for the noise figure, beat noise between the spontaneous emission and signal light becomes the major origin of the noise. The noise figure is proportional to product of the spontaneous emission factor n_{sp} and square of the excess noise factor [4]. The excess noise factor is proportional to the internal loss, facet reflectivity and gain. To improve the noise figure we should operate the SOA at high excitation level to obtain small n_{sp} close to 1. It is also effective in reducing the internal loss.

Under these guidelines, various types of SOA have been developed. Table 3.1 summarizes the performances of SOA so far reported. If we want to have large saturation power and polarization insensitive response, the bulk active layer SOA may be the most convenient available device. Table 3.1 summarizes the characteristics of various SOAs [3, 5–7].

Table 3.1   Characteristics of various SOAs

Material	Length (mm)	Current (mA)	Gain (dB)	Noise figure (dB)	Saturation power (dBm)	Reference
Bulk	0.9	500	19	7	17	[3]
MQW	1.6	500	15	5.7	16.5	[5]
Bulk	3	1500	27	7	17	[6]
Bulk	1	450	27	5.7	14.7	[7]

3.3 SOA as an Ultrafast Nonlinear Medium

Lee *et al.* investigated the electron-hole pair density dependence of the nonlinear refractive index of GaAs [8]. As the increase in the photo-excitation, absorption reduces and becomes negative for high excitation levels, correspondingly, the refractive index changes. Analysis was performed based on the Kramers–Kronig relation. This applies also to the SOA. The current injection causes optical gain and correspondingly the refractive index changes. In the adiabatic change of injection current, the refractive index change is determined by the shift of qusai-Fermi level, change in the plasma dispersion and band renormalization. If we inject an intense short optical pulse, induced carrier recombination takes place to reduce the qusai-Fermi level, and at the same time carrier heating (CH) and spectral hole burning (SHB) take place. Correspondingly, dynamical change in the optical gain and refractive index occurs. This dynamical response is composed of several time constants depending on the underlying relaxation processes.

Figure 3.2 shows the short pulse response of SOA. The output (A) is for the CW input light, and output (B) is the case when a short optical pulse was superposed on the CW input light. Operating conditions are shown in the figure. There are two responses. One is the very fast response with a time constant of the order of 1 ps. This very fast response has been investigated in detail [9, 10] and its time constant is CH cooling time and SHB recovery time. After the very fast response, there is a slow recovery of the output. The slow recovery time corresponds to the band-to-band carrier recombination lifetime and is of the order of 100 ps to 1 ns.

When we consider the use of the SOA for ultrafast signal processing, the most inconvenient feature is the slow response associated with carrier recombination. So far two methods are reported to overcome this slow response. One is the direct use of the ultrafast response by selecting only the fast response using a wavelength filter [11–17]. The other is the use of a symmetric Mach–Zehnder configuration with SOA at each arm to cancel out the slow response component, as described in Section 3.5.

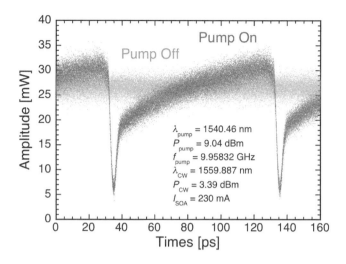

Figure 3.2 An example of the time response of SOA. Operating conditions are shown in the figure

3.4 Use of Ultrafast Response Component by Filtering

3.4.1 Theoretical Background

Optical filtering of output light to select only the ultrafast component is a powerful technique and has been utilized to enhance the modulation bandwidth of SOA-based optical switches. This method has enabled all optical wavelength conversions at 10 [11], 40 [12], 80 [13, 14], 160 [15], and 320 Gb/s [16], and optical demultiplexing from 320 to 40 Gb/s [17]. The following part of this section presents the theoretical background and experimental results on speeding up of SOA recovery time by use of an optical filter at the SOA output.

Figure 3.3 depicts an all-optical switch based on an SOA followed by an optical band-pass filter (BPF). Injected data pulses alter the gain and refractive index of the SOA resulting in amplitude modulation and chirp of the copropagating CW probe light. The leading and trailing edges of the amplitude-modulated probe light are red and blue shifted, respectively. The data pulses are rejected by the BPF, while a part of the probe light is transmitted. When the BPF is detuned towards the shorter wavelength (blue shift) with respect to the probe light, the amplitude-modulated signal recovers much faster than in the absence of the BPF [15]. A similar effect can be achieved by use of an asymmetric Mach–Zehnder interferometer, which is commonly referred to as a delayed-interference signal converter (DISC) [18, 19].

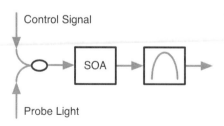

Figure 3.3 Schematic of an all-optical switch based on an SOA followed by an optical band-pass filter

Nielsen *et al.* presented a comprehensive theoretical analysis on the role of optical filtering for enhancing the modulation bandwidth of SOA-based optical switches [20,21]. In the following, the theoretical results given by Nielsen *et al.* will be briefly reviewed. It is assumed that short optical pulses are injected into the SOA as a control signal along with a strong CW probe light. Neglecting the loss in the SOA waveguide, the spatial evolution of the photon density is described by [10,22]:

$$\frac{dS_k}{dz} = \frac{\Gamma g_d (N - N_{tr})}{1 + \varepsilon (S_c + S_p)} S_k,$$
(3.3)

where S_c and S_p represent the photon densities for the control signal and CW probe light, respectively, and the subscript 'k' refers to either 'c' or 'p'. In Equation (3.3), N, N_{tr}, Γ, and g_d are the carrier density, transparency carrier density, confinement factor, and differential gain, respectively, and $\varepsilon = \varepsilon_{CH} + \varepsilon_{SHB}$ represents the nonlinear gain suppression factor including a contribution from carrier heating (CH) and spectral hole burning (SHB). Neglecting the nonlinear gain contributed from S_p, the temporal evolution of the carrier density is described by [10,22]:

$$\frac{dN}{dz} = -\frac{N - N_0}{\tau_e} - \frac{v_g g_d (N - N_{tr})}{1 + \varepsilon S_c} S_c,$$
(3.4)

where N_0 and v_g are the steady state carrier density in the absence of the control signal and group velocity, respectively. In Equation (3.4) $\tau_e = [1/\tau_{sp} + v_g g_d S_p(0)]^{-1}$ represents the effective carrier lifetime, where τ_{sp} is the spontaneous carrier lifetime.

Integration of Equations (3.3) and (3.4) yields the following differential equation for the saturated modal gain $g_m(t, z) = \ln G(t, z) = \ln[S(t, z)/S(t, 0)]$:

$$\frac{dg_m(t, z)}{dt} = \frac{1}{1 + \varepsilon \exp[g_m(t, z)]S(t, 0)} \times \left[-\{\exp[g_m(t, z)] - 1\} \left(v_g g_d + \varepsilon \frac{d}{dt} \right) S(t, 0) \right.$$
$$\left. - \frac{g_m(t, z) - g_{m,0}(z)}{\tau_e} + \varepsilon \{\exp[g_m(t, z)] - 1\}S(t, 0) \right]$$
(3.5)

where $S(t, 0) = S_c(t, 0) + S_p(0)$ and $g_{m,0}(z) = \Gamma g_d (N_0 - N_{tr})z$ is the unsaturated modal gain. The solution of Equation (3.5) provides the output power of the probe light $S_p(t, z) = \exp[g_m(t, z)]S_p(0)$. By setting $\varepsilon = 0$, which corresponds to the exclusion of ultrafast carrier dynamics due to CH and SHB, Equation (3.5) reduces to the linear model of Agrawal and Olsson [23].

The total phase shift $\Phi(t, z)$ is given by the sum of linear and carrier heating contribution and is described by

$$\Phi(t, z) = \Phi_N(t, z) + \Phi_{CH}(t, z)$$

$$= -\frac{1}{2} \alpha_N [g_m(t, z) - g_{m,0}(z)]$$

$$+ \frac{1}{2} \alpha_{CH} \varepsilon_{CH} \{\exp[g_m(t, z)] - 1\}S_p(0)$$
(3.6)

The contribution of spectral hole burning can be neglected when the wavelength of the control signal is close to the SOA gain peak. In Equation (3.6) α_N is the line width enhancement factor and α_{CH} is the phase-amplitude coupling coefficient due to electron temperature change.

From the above solutions, the probe light electric field, $E_p(t, L)$, at the SOA output can be expressed by:

$$E_p(t, L) = \sqrt{G(t, L)\xi S_p(0)} \exp[i\Phi(t, L)] \tag{3.7}$$

where L is the SOA length and ξ is the conversion factor from photon density to optical power. The action of an optical filter at the SOA output can be described in a frequency domain using the filter transfer function $H_F(\omega)$ as:

$$E_F(L) = F^{-1}\{H_F(\omega) F[E_p(t, L)]\} \tag{3.8}$$

where $F[\bullet]$ and $F^{-1}[\bullet]$ represent the Fourier and inverse Fourier transforms, respectively. In the case of a Gaussian BPF $H_F(\omega)$ is given by:

$$H_{Gauss}(\omega) = \exp\left[-2\ln 2 \left(\frac{\omega - \omega_c}{\Delta\omega_{1/2}}\right)^2\right], \tag{3.9}$$

where ω_c and $\Delta\omega_{1/2}$ denote the center frequency and full width at half maximum (FWHM) of the filter, respectively. The transfer function of an asymmetric Mach–Zehnder interferometer is given by

$$H_{AMZ}(\omega) = \frac{1 + \exp[i(\omega\tau + \phi_b)]}{2}, \tag{3.10}$$

where τ and ϕ_b denote the differential delay time and phase bias of the interferometer, respectively. Equations (3.3) – (3.10) can be solved numerically.

Figure 3.4 shows the results of simulation and experiment for the output waveforms of an SOA optical switch followed by a Gaussian BPF with FWHM of 0.9 nm [21]. The left and middle columns represent results of simulation excluding and including ultrafast carrier dynamics due to CH and SHB, while the right-hand column shows the experimental results. The top row corresponds to the SOA output, while the second, third, and bottom rows represent waveforms at the BPF output with the BPF detuning of -0.7, -1.3, and -1.5 nm, respectively. The control signal is a 100-Gb/s '1111000000' data stream with pulse duration and energy of 1.8 ps and 133 fJ, respectively. From the comparison of waveforms in Figure 3.4, it can be seen that ultrafast carrier dynamics reduce patterning effects, and that the simulated results, including ultrafast carrier dynamics, agree better with experimental results. The effect of the detuned BPF can be clearly displayed in Figure 3.4 for enhancing the modulation bandwidth. It should be noted that for small negative detuning (-0.7 nm) the signal polarity is inverted as a result of the enhancement of the negatively detuned rising edge, which can be interpreted as BPF-assisted cross gain modulation. For large negative detuning (-1.3, -1.5 nm) the signal polarity is preserved because the BPF suppresses the carrier component of the probe light, which makes cross phase modulation dominant. Similar behavior was observed for a positively detuned BPF [24].

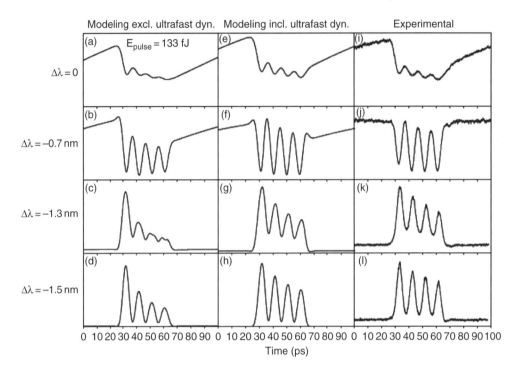

Figure 3.4   Output waveforms for an SOA optical switch followed by a 0.9 nm Gaussian BPF [21]. (Reproduced by permission of OSA)

3.4.2 Signal Processing Using the Fast Response Component of SOA

This section describes applications of SOA optical switches with optical BPFs to all-optical signal processing. Table 3.2 summarizes results of experiments that employed SOA-based optical switches assisted by optical filtering. Wavelength conversion has been demonstrated for bit rates up to 320 Gb/s. It can be seen that both the full width at half maximum (FWHM) and detuning increase for higher bit rate signals due to wider spectral bandwidth. Since the polarity of wavelength-converted RZ signal is inverted by cross-gain modulation, notch filters that suppress the DC component were used after BPFs to obtain noninverted data signals. Notch filters were realized using polarization interferometers [13, 15, 16]. Non-inverted wavelength conversion was also demonstrated by use of a single BPF with a small FWHM, which served to suppress the DC component as well as to select the blue-shifted sideband [12]. Numerical optimization of the transfer function of filters after SOAs was studied using genetic algorithms, which clarified the dependence of the signal quality on the bandwidth and detuning of filters [25]. Using a similar principle, all-optical demultiplexing was demonstrated in which 160- and 320-Gb/s optical time-division-multiplexed (OTDM) data streams were demultiplexed into 40-Gb/s base-rate channels. In this experiment, a continuous wave (CW) probe light used in wavelength conversion was replaced with a 40-GHz clock signal. One of the advantages of SOA optical switches with optical BPFs is their simple configuration enabling photonic integration [14].

Table 3.2 Optical signal processing experiments using SOA-based optical switches assisted by optical filtering

Function	Bit rate [Gb/s]	Control signal	Filter	Remarks	Reference
Wavelength conversion	10 (2^{31}–1)	RZ		20,000 km transmission	[11]
Wavelength conversion	40 (2^{31}–1)	7 ps RZ 3 fJ/pulse	0.22 nm FWHM −0.5 nm detuning	1.7 dB penalty	[12]
Wavelength conversion	80 (2^7–1)	2.5 ps RZ 21 fJ/pulse	0.72 nm FWHM −0.5 nm detuning	3 dB penalty	[13]
Wavelength conversion	80 (2^7–1)	2 ps RZ	AWG (20 nm FSR) 1 nm FWHM	Monolithic device 10 dB penalty	[14]
Wavelength conversion	160 (2^7–1)	1.9 ps RZ 15 fJ/pulse	1.4 nm FWHM −1.23 nm detuning	2.5 dB penalty	[15]
Wavelength conversion	320 (2^7–1)	1.7 ps RZ 5.5 fJ/pulse	2.7 nm FWHM −2.55 nm detuning	10 dB penalty	[16]
OTDM demultiplexing	320/40 (2^7–1)	0.8 ps RZ 9.8 fJ/pulse	1.5 nm FWHM		[17]

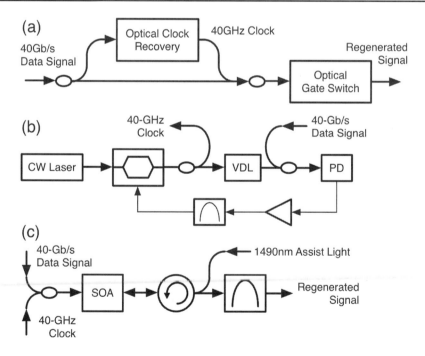

Figure 3.5 Schematics of (a) 40-Gb/s optical 3R regenerator, (b) optical clock recovery, and (c) SOA-based optical gate switch

In the following, an optical 3R (re-amplification, re-timing, re-shaping) regeneration experiment is described as an example of all-optical signal processing. Figure 3.5(a) is a diagram of a 40-Gb/s optical 3R regenerator consisting of optical clock recovery and SOA optical switch. First, a 40-GHz optical clock signal is recovered from the incoming 40-Gb/s data signal, which is then input to an optical gate switch that is controlled by the incoming data signal.

Figure 3.5(b) shows a schematic of an optical clock recovery employing a 40-GHz optoelectronic oscillator (OEO) [26–28]. The OEO is constructed from CW laser, LiNbO$_3$ intensity modulator, variable optical delay line (VDL), photodiode (PD), RF amplifiers, and electrical band-pass filter (BPF). A continuous-wave (CW) light at 1556.0 nm is input to the modulator and is converted to an electrical signal by the PD. The PD output signal is amplified and is fed back to the modulator through the BPF, whose center frequency and Q-factor are 39.81291 GHz and 961, respectively. In the absence of incoming data signals, the OEO supports self-sustained oscillation at the frequency determined by the BPF and transit time of the feedback loop. By injecting 40-Gb/s optical data signals directly into the PD, the OEO is forced to synchronize with the data signals, because the clock component in the data signals is selectively amplified in the OEO feedback loop, resulting in injection locking of the OEO. The optical clock

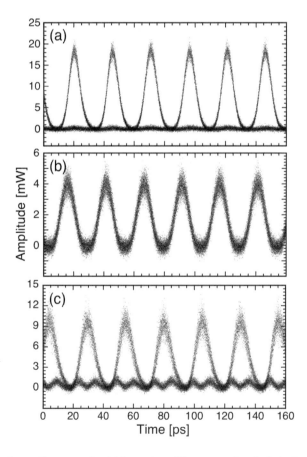

Figure 3.6 Optical sampling traces for (a) input data, (b) recovered optical clock, and (c) regenerated data signals

is extracted just after the modulator using a directional coupler (DC), whereas the 40-Gb/s data signal is directly coupled to the PD. Figure 3.5(c) shows schematic of an optical gate switch consisting of SOA, BPF with 1-nm FWHM, and circulator. The SOA is a commercial product designed as a pre-amplifier and has small signal gain and saturation output power of 22.1 dB and 13.3 dBm at 1550 nm, respectively. 40-Gb/s data and 40-GHz clock signals are combined by a directional coupler and are input to the SOA. A counter-propagating assist light at 1490 nm is injected for reducing the SOA recovery time. For measuring an bit error rate (BER) the regenerated data signal is detected by a 40-Gb/s photoreceiver and is demultiplexed into 10-Gb/s signals using an electronic 1:4 demultiplexer.

39.81312-Gb/s return-to-zero (RZ) optical data signal is generated from an actively mode-locked fiber ring laser (MLFRL) with a repetition frequency of 9.95328 GHz. The wavelength and pulse width of the MLFRL output are 1547.6 nm and 9.1 ps, respectively. The MLFRL output is encoded by a $2^7 - 1$ pseudorandom bit sequence (PRBS), and 39.81312-Gb/s data streams are generated using fiber-optic OTDM multiplexers. Figures 3.6(a) and (b) show optical sampling traces for the input data and recovered clock signals, respectively. The pulse width of the clock signal is 11.0 ps. Figure 3.6(c) shows the optical sampling trace for the regenerated data signal. The detuning of the BPF is -0.6 nm. The average power inputs to the SOA are -4.92 and 8.58 dBm for the clock and data signals, respectively. Since a clock signal is used as a probe instead of a CW light, it is not possible to obtain a noninverted data signal by a notch filter. The observed undulation in the baseline of the regenerated signal is caused by the pulse width of the input data signal being smaller than that of the clock signal.

Figure 3.7 shows BER for the demultiplexed 10-Gb/s signals plotted as a function of the received power. Curve A (squares) corresponds to back-to-back operation with the receiver

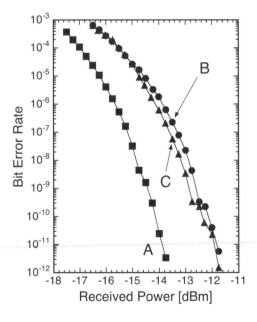

Figure 3.7   BER curves plotted as a function of the received power. Curve A (square) corresponds to back-to-back operation, while curves B (circle) and C (triangle) are obtained by 3R regeneration and wavelength conversion, respectively

sensitivity of $-14.25\,\text{dBm}$ at $\text{BER} = 10^{-9}$. Curve B (circles) represents the results for 3R regeneration corresponding to the trace in Figure 3.6(c). Error free operation is achieved with the power penalty of 1.7 dB. For comparison, curve C (triangles) in Figure 3.7 represents the result for wavelength conversion, in which the clock signal is replaced with the CW light and a notch filter is added after the BPF in Figure 3.5. The power penalty for wavelength conversion is 1.5 dB, which is slightly better than that obtained in 3R regeneration. Operation for higher bit rate signals can be achieved by properly designing the FWHM and detuning of the BPF.

3.5 Symmetric Mach–Zehnder (SMZ) All-Optical Gate

3.5.1 Fundamentals of the SMZ All-Optical Gate

In this section, we describe all-optical signal processing using SOAs in the Symmetric Mach–Zehnder (SMZ) configuration. In the SMZ gate, nonlinear optical effects associated with the carrier density change in SOAs are utilized. The slow response component due to the band-to-band electron recombination is canceled out using a differential phase modulation scheme.

Figure 3.8(a) shows the configuration of the SMZ gate [29]. SOAs used as nonlinear waveguides are placed in both arms of a Mach–Zehnder interferometer. The control light governs the dynamics of nonlinear optical effects and the probe light experiences the nonlinear optical effects. The control light can be either return-to-zero (RZ) pulses or non-return-to-zero (NRZ) light. In Figure 3.8, the SMZ gate with the control light of RZ pulses and the probe light of RZ pulses is shown. In SOAs, the control pulse induces carrier depletion and thus modulates the gain and phase of the probe light. These phenomena are called cross-gain modulation (XGM) and cross-phase modulation (XPM), respectively. These nonlinear optical effects induced through the carrier density change in semiconductors, are generally highly efficient, which means that device operation can be realized in a compact size and with low power control light. In addition to this, the control light, which depletes carriers, is amplified in SOA. This further lowers the required power of input control light. The major obstacle in using XGM and XPM with the carrier density change in SOAs is its slow relaxation time. Typical relaxation times are in the range of 100 ps to 10 ns. In the SMZ gate, however, ultrafast gating is achieved by a so-called differential phase modulation scheme. By exciting the two SOAs, arranged symmetrically on both arms of the Mach–Zehnder interferometer, with an appropriate time delay ΔT, the effect of the slow relaxation is canceled out.

The dynamics of carrier density in SOAs and associated nonlinear optical effects can be analyzed by rate equations [23]:

$$\frac{dN}{dt} = J - \frac{N}{\tau} - g_d(N - N_{tr})\frac{S_c}{\hbar\omega_c} - g_d(N - N_{tr})\frac{S_p}{\hbar\omega_p} \qquad (3.11)$$

$$\frac{dS_c}{dz} = \Gamma g_d(N - N_{tr})S_c - aS_c \qquad (3.12)$$

$$\frac{dS_p}{dz} = \Gamma g_d(N - N_{tr})S_p - aS_p \qquad (3.13)$$

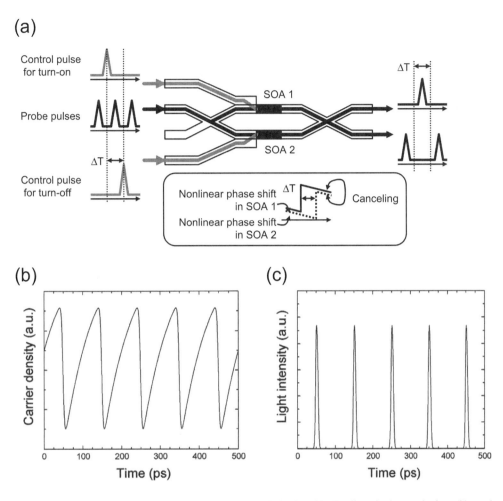

Figure 3.8 (a) Configuration of a symmetric Mach–Zehnder (SMZ) all-optical gate device; (b) total carrier number in SOA; (c) output of SMZ gate

where N is the carrier density, J is the rate of carrier injection through bias current, N_{tr} is the transparency carrier density, S_c is the control light power, S_p is the probe light power, ω_c is the control light frequency, ω_p is the probe light frequency, g_d is the differential gain, τ is the carrier lifetime, Γ is the light confinement factor, and a is the optical loss coefficient including absorption and scattering. The modulation in the intensity of the probe light is derived from the above equations. The modulation in the phase (the nonlinear phase shift) is given by the combination of the alpha-parameter (which is sometimes called the line-width enhancement factor) and the gain:

$$\frac{d\phi}{dz} = -\frac{1}{2}\alpha_N \Gamma g_d (N - N_{tr})\tag{3.14}$$

where ϕ is the phase of the probe light and α_N is the alpha-parameter. α_N actually depends on material, wavelength, and other parameters but the typical value is about 3 to 6. As can be seen, the phase shift is proportional to the total carrier number in SOA (integral of N along the entire length of SOA). This is a good approximation for the estimation of phase shift. In a general analysis of SOAs, the influence of amplified spontaneous emission (ASE) is also important. However, in the situations considered here for all-optical signal processing, control light highly depletes carriers in SOAs and so, neglecting the effect of ASE, still gives good approximation to the carrier dynamics of SOAs.

Figure 3.8(b) shows the change in the total carrier number in SOAs and Figure 3.8(c) shows the gating window with the SMZ gate calculated using the above rate equations. Here the gating window is represented as the intensity modulation on the CW probe light. Control pulses with a duration of 6 ps at a repetition rate of 10 GHz, cause the change in the total carrier number or the carrier density in SOAs. The depletion in the carrier density is fast following the short-duration control pulse, but the recovery in the carrier density is slow with a time constant of 250 ps. However, by exciting both arms with an interval ΔT of 6 ps, the gating window for the probe light opens only within 6 ps. In other time regions, destructive interference for the probe light at the output of the Mach–Zehnder interferometer is maintained even though the carrier density in the SOAs on both arms is slowly recovered. Both the rising and the falling of the gate window are defined by the control pulse duration.

The mechanism of canceling out the slow relaxation of the carrier density change can be implemented in other configurations. Using one SOA and the interference of orthogonally polarized probe light beams is another method, called the polarization-discriminating symmetric-Mach–Zehnder (PD-SMZ) all-optical gate [30] or an ultrafast nonlinear interferometer (UNI) [31]. Figure 3.9 shows a schematic of the PD-SMZ gate, which consists of an SOA as a nonlinear waveguide, birefringent crystals (BC1-2), and a polarizer (PL). Here we consider the output state of three probe pulses. Each probe pulse is split into a pair of orthogonally polarized and temporally displaced components in BC1. The time delay ΔT between the two components is determined by the difference in group delays for ordinary and extraordinary light in BC1. After BC1, the two components of the probe pulse are launched into the SOA along with the control pulses. In the SOA, the probe pulse experiences the nonlinear phase shift induced by the control pulses. In Figure 3.9, probe pulse components indicated by 3a, 2b, and 3b experience a nonlinear phase shift. After the SOA, a pair of the orthogonally polarized and temporally displaced components of the probe pulse is recombined in BC2. The polarization state of the recombined probe light is determined by the phase difference between the two orthogonally polarized components. Only within the time window ΔT, is the polarization state different from the initial state formed. In Figure 3.9, the polarization of the probe pulse

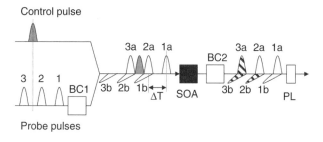

Figure 3.9 Configuration of polarization discriminating (PD-)-SMZ gate. BC1-2: birefringent crystals; PL: polarizer

consisting of components indicated by 2a and 2b is different from others. With this mechanism, a gate window unrestricted by the relaxation of induced nonlinear phase shift is obtained. In the setup of the PD-SMZ gate, two orthogonally polarized probe light beams are propagated through the same path and thus interfere stably, which greatly facilitates experiments using discrete optical components with long optical paths.

When the probe light is CW light, the setup can be simplified [18, 32]. The combination of one SOA and a delay interferometer, in other words, an asymmetric Mach–Zehnder filter, is useful for both integrated devices and discrete component setup experiments [18]. A Sagnac interferometer-based configuration is also proposed [33]. In fact, there are various configurations that utilize the mechanism of canceling out the relaxation tail of carrier density change. One of the important points for categorizing these configurations is whether or not the control light and the probe light are co-propagated or counter-propagated through the SOAs used as nonlinear waveguides. When the probe light is counter propagated with the control light, the gate window shape is influenced by the time during which the control light propagates thorough the entire length of the SOA. As discussed in Refs [34–36], the SOA length of $500\,\mu m$ causes the gate window edge to broaden by a few ps.

3.5.2 Technology of Integrating Optical Circuits for an SMZ All-Optical Gate

The technology of integrating optical circuits is also important for developing practical SMZ all-optical gates. SOAs used as nonlinear waveguides should be incorporated into Mach–Zehnder interferometers consisting of low loss, passive waveguides. The integration of active and passive waveguides for SMZ gates has been developed both in a hybrid manner and in a monolithic manner. Figure 3.10(a) shows a hybrid integrated device where SOAs are mounted on a silica-based planar lightwave circuit (PLC) [35, 36]. The SOAs [39] consisted of a bulk InGaAsP core and InP claddings with integrated spot-size converters (SSCs). To reduce the facet reflectivity, window regions and angled facets were introduced, and anti-reflection coating was applied. Silica-based waveguides forming a Mach–Zehnder interferometer were fabricated on a silicon wafer by atmospheric-pressure chemical-vapor deposition (AP-CVD). Several methods of mounting SOAs on silica-based PLC platforms have been reported, such as flip-chip bonding SOAs on PLC platforms [37, 38, 40, 41] inserting silicon blocks carrying SOAs into gaps formed on PLC platforms [42], and mounting SOAs followed by the subsequent mounting of PLC parts [43, 44]. In Figure 3.10, the central area of the Mach–Zehnder interferometer is etched and then striped solder bumps are formed. An SOA-array chip is mounted on this area in a self-aligned manner, using the solder bumps on the silicon substrate together with solder wettable pads formed on the SOA array chip [40]. On both arms, static phase controllers are placed. These static phase controllers are based on a thermo-optical effect induced by heating a portion of the silica-based waveguides. With this mechanism, the phase of probe light propagated through the two arms of the Mach–Zehnder interferometer can be adjusted. The optical fiber arrays are assembled using V-grooves formed on the silicon substrate. Figure 3.10(b) shows a packaged SMZ gate module and electronic bias circuits. Electronic circuits supply injection current to SOAs and heater current to static phase controllers. Because high-speed signal processing is done within the optical domain with the SMZ gate, there is no need to assemble high-speed electronic circuits.

Various methods of monolithically integrating active and passive waveguides for SMZ gates have also been reported [45–52]. SOAs are butt jointed to passive waveguides with a repeating

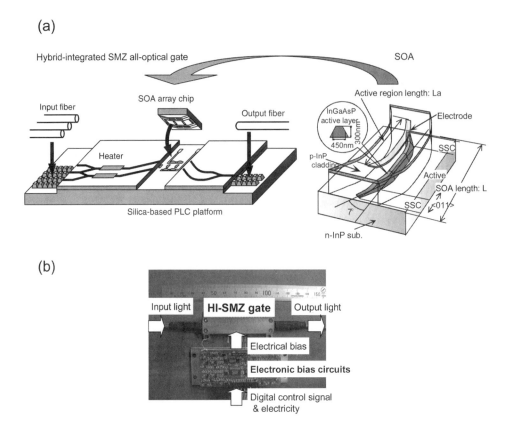

Figure 3.10 (a) Structure of SOA and hybrid-integrated SMZ gate. (b) Photograph of a packaged SMZ gate module and electronic bias circuit

growth process, formed as upper layer waveguides on passive waveguides, or simultaneously formed with selective area growth technology. Monolithic integration technology has now been developed to the extent that SMZ gates are integrated together with other active devices, such as input and output optical amplifiers and fixed or tunable diode lasers. Significant reduction in the footprint of Mach–Zehnder optical circuits has been demonstrated using photonic-crystal waveguide technology [53]. The passive waveguide portion can be miniaturized with sharp bending and a compact splitter/combiner, and the active waveguide portion can be miniaturized using slow light structure.

3.5.3 Optical Demultiplexing

In this section, the optical demultiplexing achieved by an SMZ gate is described as a typical example of operation where a very narrow optical gating window is required. If we consider the receiver for a 160 Gb/s serial optical signal, we should extract tributaries at lower bit rates or convert serial to lower bit rate parallel in the optical domain before optical-to-electrical

(OE) conversion. This is essential because direct OE conversion using a photodetector and subsequent electronic processing at 160 Gb/s is quite difficult.

Optical demultiplexing from 160 Gb/s to 10 Gb/s can be carried out by the SMZ gates [54]. The operation scheme for optical demultiplexing with the SMZ gate is shown in Figure 3.11. In this case, a 6.25-ps gating window is opened by exciting the SMZ gate with control pulses at a repetition rate of 10 GHz. The signal pulses conveying 160 Gb/s data with on–off keying is also input into the SMZ gate as probe light. One in every 16 optical signal pulses is extracted with the gating window and thus 10 Gb/s optical signal is extracted. Figure 3.12 shows a typical experimental setup for demonstrating optical demultiplexing using the hybrid-integrated SMZ (HI-SMZ) gate. Signal pulses of 168 Gb/s were generated by modulating the output of an actively mode-locked fiber laser at a repetition rate of 10.5 GHz and then multiplexing these modulated pulses. Optical pulses at 10.5 GHz from another fiber laser synchronized to the first fiber laser were used as control pulses. The pulse energies of the control and signal pulses coupled to each SOA were about 200 and 1 fJ, respectively. Typical bias currents applied to SOAs were 100 mA. Using a wavelength filter at the output of the HI-SMZ gate, the demultiplexed signal pulses are extracted and the control pulses are blocked.

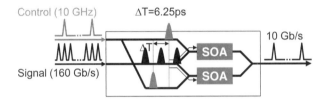

Figure 3.11   Operation scheme for optical demultiplexing with SMZ gate

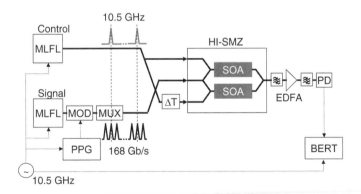

Figure 3.12   Typical experimental setup for demonstrating optical demultiplexing with an SMZ gate. MLFL, mode-locked fiber laser; MOD, modulator; MUX, multiplexer; EDFA, erbium doped fiber amplifier; PD, photodetector; PG, pulse pattern generator; BERT, bit error rate tester

Figure 3.13 shows the measured results for optical demultiplexing from 168 Gb/s to 10.5 Gb/s. Signal pulses of 168 Gb/s with a duration of 1.2 ps at a wavelength of 1560 nm and 10.5 GHz control pulses with a duration of 1.9 ps at a wavelength of 1545 nm were used in this measure-ment. As mentioned above, there are no photodetectors having speed sufficient for 160 Gb/s, thus waveform measurement also needs optical domain processing. Here the eye diagrams of input 168 Gb/s signals and output 10.5 Gb/s signals shown in Figures 3.13(a) and (b) were measured using an optical sampling oscilloscope. The results of measuring averaged output waveform with high signal-to-noise ratio shown in Figure 3.13(c) indicate that the intensity contrast between demultiplexed pulses and other pulses are about 23 dB. As shown in Figure 3.13(d), a power penalty of about 1.5 dB at a bit error rate (BER) of 10^{-9} was achieved for a pseudo-random bit sequence (PRBS) length of $2^{31} - 1$. Other experiments using a 1.3-μm-band, SOA-based SMZ gate or monolithically integrated SMZ gate have also successfully demonstrated optical demultiplexing from 160 Gb/s to 10 Gb/s [55–57]. Demonstration in the transmission setup has also been performed together with using clock recovery [58, 59].

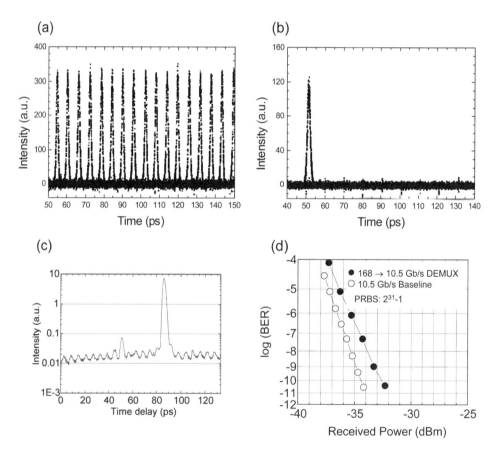

Figure 3.13 Experimental results for optical demultiplexing from 168 Gb/s to 10.5 Gb/s with an SMZ gate: (a) input signal pulse eye diagram; (b) output (demultiplexed) signal pulse eye diagram; (c) averaged output waveform for evaluating contrast; (d) measured bit error rates of output signal

An approach to improving switch performance such as a signal-to-noise ratio or operation speed could be to use shorter durations for control and signal pulses. However, the use of
extremely shortened pulses causes difficulties in generating and handling pulses. It is important
to quantitatively evaluate the effect of the durations of the control and signal pulses. However,
contrary to the above intuition, experimental results on 168-Gb/s demultiplexing using control pulse durations of 1.6 and 3.5 ps showed little difference in the power penalty [60]. The
influence of control and signal pulse duration can be discussed by considering pulse duration
dependence of cross talk between the channel to be demultiplexed and nearest neighbor channels. This cross talk at the optical demultiplexer leads to fluctuation in the output voltage of
the subsequent OE converter. For example, the output voltage corresponding to '1' fluctuates
depending on the input patterns of '111', '110', '011', and '010'. Here the three serial bits
represent the channel to be demultiplexed and the nearest-neighbor channels. A Q-factor is
defined by connecting the mean and the standard deviation of the OE converter output with the
input patterns. Figure 3.14 shows calculated results for the case of 168-Gbit/s demultiplexing
with a 6-ps switching window. The dependence on the control pulse duration and on the signal
pulse duration are quite different. When the signal pulse is rather broad, shortening the control
pulse does not improve the Q-factor. In contrast, shortening the signal pulse improves the
Q-factor rather independently of the control pulse duration.

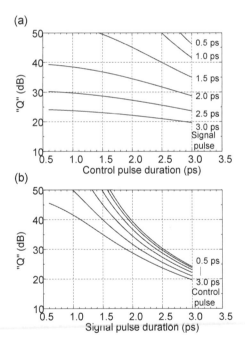

Figure 3.14 Dependence of cross talk caused by nearest-neighbor symbols on control pulse duration
and signal pulse duration

The weak dependence of the cross talk on the control pulse is attributed to the sinusoidal
nature of Mach–Zehnder interferometer output. The nonlinear phase shift experienced by the

signal light transmitted through an SOA is considered to be proportional to the amount of carrier depletion induced by the control pulse, but the transmission provided by the SMZ gate sinusoidally depends on the nonlinear phase change. Thus, the rising and falling edges of the switching window opened by the SMZ gate are steeper than what would be expected from the control pulse duration. There is no need to use extremely shortened control pulses, which is advantageous for not only generating and handling pulses but also minimizing the effect of carrier heating in SOA nonlinearity. When a sub-picosecond control pulse is used to excite SOAs, the effect of carrier heating increases [61]. Carrier heating reduces the amount of carrier depletion. As a result, the magnitude of a nonlinear phase shift associated with carrier depletion decreases. Carrier heating also adds another component to the nonlinear phase shift. This component of the nonlinear phase shift has a relaxation time of about 1 ps. In the SMZ switch, which utilizes canceling between the relaxation tails of the nonlinear phase shifts induced in the SOAs of the two arms, the nonlinear phase shift associated with carrier heating causes degradation in the extinction ratio emerging after the switching window.

Considering the above guideline, optical demultiplexing from 336 Gb/s to 10.5 Gb/s [62] and from 672 Gb/s to 10.5 Gb/s has also been demonstrated. Figure 3.15 shows the eye diagrams of 672 Gb/s input signal pulses and 10.5 Gb/s output signal pulses. In this experiment, the signal pulse with a duration of 0.5 ps at a wavelength of 1555 nm and a control pulse with a duration of 0.8 ps at a wavelength of 1530 nm were used. Compared with the gating window of 1.5 ps for demultiplexing from 672 Gb/s, the signal pulse duration was set to about one third and the control pulse duration was set to about one half.

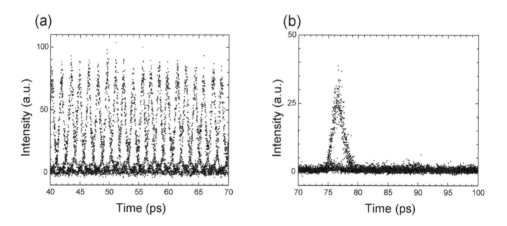

Figure 3.15 Results for optical demultiplexing from 672 Gb/s to 10.5 Gb/s: (a) input signal pulse eye diagram; (b) output signal pulse eye diagram

In the above, the SMZ gates driven by 10-GHz control pulses are demonstrated. Optical demultiplexing using higher repetition rate control pulses is also possible using SMZ gates. By using control pulses at a repetition rate of 42 GHz, optical demultiplexing from 168 Gb/s to 42 Gb/s and from 336 Gb/s to 42 Gb/s has also been demonstrated. Figure 3.16 summarizes the measured result when 168-Gb/s signal pulses are demultiplexed to 42 Gb/s.

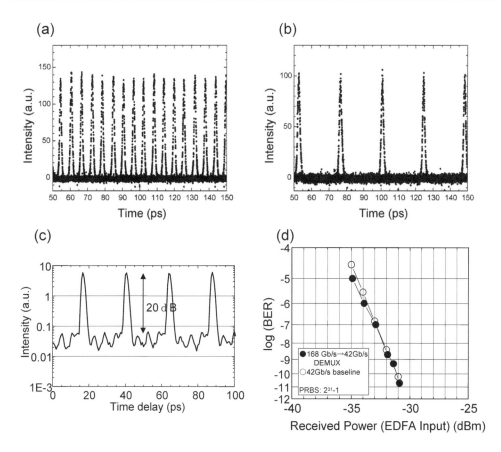

Figure 3.16 Results for optical demultiplexing from 168 Gb/s to 42 Gb/s: (a) input signal pulse eye diagram; (b) output signal pulse eye diagram; (c) averaged output waveform for evaluating contrast; (d) measured bit error rates of output signal

3.5.4 Wavelength Conversion and Signal Regeneration

The use of SMZ all-optical gates is also expected in applications to wavelength conversion and signal regeneration. The operation schemes with SMZ gates are shown in Figure 3.17. In the 3R (re-timing, re-shaping, and re-amplifying) regeneration shown in Figure 3.17(a), data-modulated pulses excite the SMZ gate and open a switching window for clean clock pulses, usually with a different wavelength. Consequently, re-timed and re-shaped data pulses, newly created by modulating the clock pulses, are output from the SMZ gate. When the re-timing operation is not needed, in other words, functions such as wavelength conversion or 2R regeneration are needed as shown in Figure 3.17(b), CW light instead of clean clock pulses is injected into the SMZ gate and modulated all-optically.

In applying SMZ gates to wavelength conversion and signal regeneration, SOAs are excited by high-repetition, data-modulated pulses. An important issue is to minimize the pattern effect on the excited nonlinear phase shift in SOAs. A well-known method for coping with the pattern effect is the use of a holding beam, which effectively reduces the carrier lifetime in

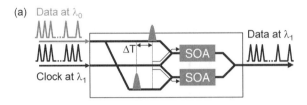

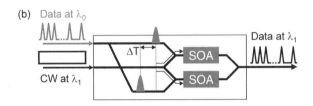

Figure 3.17 Operation scheme FOR all-optical wavelength conversion and signal regeneration with an SMZ gate: (a) 3R operation; (b) 2R operation

SOAs [63, 64]. Although in some cases the holding beam is separately prepared, probe light with stationary power, that is, clock pulses in 3R regeneration and CW light in 2R regeneration, can be used as a holding beam by setting its power higher. This can be understood by slightly modifying the rate equation of the carrier density (Equation 3.11). As S_p in Equation (3.11) does not temporally change, the fourth term of the right-hand side of Equation (3.11) plays a role as a recovery term like the second term. By combining the second term and the fourth term, the effective carrier lifetime τ_{eff} is represented by:

$$\frac{1}{\tau_{eff}} = \frac{1}{\tau} + \frac{1}{\hbar\omega_p/g_d S_p}. \tag{3.15}$$

The drawback of this method is that the intense clock pulses or the CW light consume carriers and thus reduce the refractive index change induced by the data pulses. Under the limited amount of carrier injection into SOAs, maximizing a nonlinear phase shift and minimizing the pattern effect would be compromised.

To increase carrier injection, longer SOAs are useful. The effect of SOA length has also been intensively investigated. Under the conditions of a lower optical power being injected, it has been pointed out that the bandwidth of cross-gain modulation is improved in longer SOAs [65,66]. This effect is attributed to effective reduction in the carrier lifetime in the rear part of the SOA. However, characteristics relevant to the SMZ gate for signal regeneration and wavelength conversion require that intense stationary light be injected. Experimental comparison in the induced nonlinear phase shift has been carried out for different SOAs, which were identical in cross-sectional structure but had different current-injected active region lengths (L_a), under the condition typically used for signal regeneration and wavelength conversion at 42 Gb/s [39,67]. The cross-section of the InGaAsP active region was 0.45 μm wide and 0.3 μm thick. The lengths of the SSC region and the window region were 225 and 25 μm, respectively. The nonlinear phase shifts induced in SOAs with $L_a = 0.55$ mm and $L_a = 1.67$ mm were measured under 1553-nm CW light with an average power of 8 dBm and 1564 nm, injecting 42-GHz pulses with an energy of 50 fJ (corresponding to 0 dBm for 42-Gb/s data pulses with a mark ratio of $1/2$). After propagating through the SOA, the optical spectrum of the CW light was broadened, reflecting cross-phase modulation and cross-gain modulation induced by 42-GHz, 3.5-ps pulses. The resultant optical spectrum was composed of discrete components at an interval of 42 GHz. The magnitude of the nonlinear phase shift was estimated from the intensity ratio between the main component and the first higher-frequency side component of the optical spectrum [68]. Figure 3.18 shows the measured nonlinear phase shift as a function of the injection current. In the longer SOA, the nonlinear phase shift reached a higher level by injecting higher current. When both the nonlinear phase shift and the injection current are normalized to the current-injected active region length, the curves are approximately equal for both SOAs. At a normalized injection current of 360 mA/mm, a normalized nonlinear phase shift of about 0.4 π/mm was induced for both SOAs. This indicates that the nonlinear phase shift induced in the entire SOA length is nearly proportional to the active region length under the injection of the same current density.

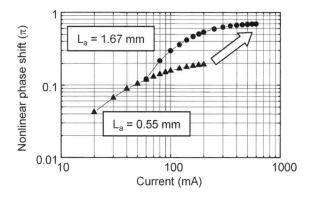

Figure 3.18 Comparison of nonlinear phase shifts induced in SOAs with different lengths

Effectively reducing the carrier lifetime in SOAs by using the so-called holding beam effect explained above or by optimizing SOA structure is useful for high bit rate operation. In fact, all-optical wavelength conversion at 40 Gb/s using XPM in SOA without differential modulation scheme has been demonstrated [69–71]. However, using the mechanism of the SMZ gate in addition to the effective reduction in carrier lifetime provides a more efficient method to reduce the pattern effect in the output waveform. Figure 3.19 shows the simulated results of total carrier number in SOA and the output of the SMZ gate. Eliminating the pattern effect in the carrier density change as shown in Figure 3.19(a) needs a fairly high intensity holding beam. On the contrary, even though the pattern effect remains in the change in the total carrier number in SOAs as shown in Figure 3.19(b), the output of the SMZ gate exhibits a high quality waveform as shown in Figure 3.19(c). In the SMZ switch, the mechanism of canceling out the relaxation tail of carrier density change minimizes the influence of the previous bit on the next bit in the output waveform, which also means a reduction in the pattern effect.

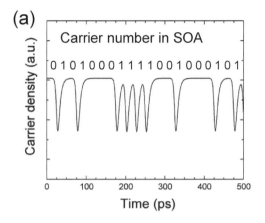

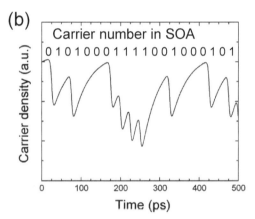

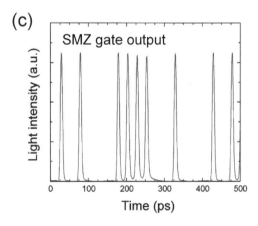

Figure 3.19   Total carrier number in SOA and output of SMZ gate

Wavelength conversion and signal regeneration based on SMZ gates are being intensively studied [72–79]. Figure 3.20 shows experimental results on wavelength conversion at 42 Gb/s using HI-SMZ gates in which SOAs with $L_a = 1.67$ mm are mounted. Typical bias currents applied to SOAs are about 300 mA. RZ data pulses with a duration of 4 ps at a wavelength of 1558 nm are input and thus play a role as control light in the SOAs. CW light at 1544 nm is also input as probe light. The average power of the data pulses and the CW light coupled to the SOA in the HI-SMZ gate is about − 1 dBm and + 5 dBm, respectively. Thus CW light is called the probe light but its average power is higher than that of the data pulses. In SOAs, this CW light is effectively a holding beam to suppress the pattern effect in the carrier density change. By setting the interval of exciting both arms of the HI-SMZ gate at 6 ps in this case, wavelength-converted 42 Gb/s data pulses at 1544 nm with a duration of 6 ps are output. Measured eye diagrams and bit error rates shown in Figure 3.20 prove high quality output. Clock pulses

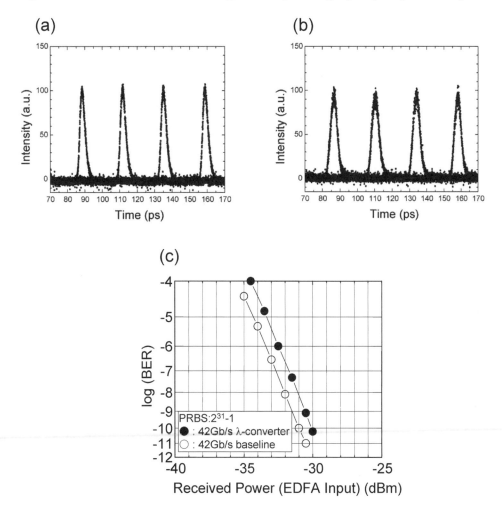

Figure 3.20 Experimental results for 42 Gb/s wavelength conversion: (a) input data eye diagram; (b) output (wavelength-converted) data eye diagram

instead of CW light can be used with nearly the same average power conditions [78]. Within the wavelength range of 30 nm, little degradation in the results of bit error rate measurement has been observed [78]. Wavelength conversion at 168 Gb/s has also been achieved. Figure 3.21 shows the eye diagrams of input data pulses and output data pulses. Error-free operation has been demonstrated in several separate experiments using the mechanism based on the SMZ gates.

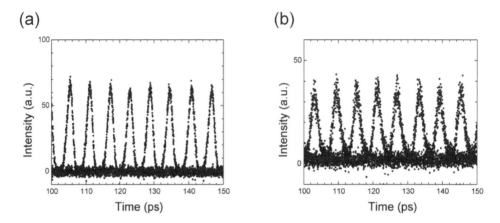

Figure 3.21 Experimental results for 168 Gb/s wavelength conversion: (a) input data eye diagram; (b) output (wavelength-converted) data eye diagram

When the pattern effect is sufficiently suppressed, we can expect reshaping properties in the operation of the SMZ gate. This is very useful for signal regeneration in the optical domain, which should be cost-effective particularly at high bit rates. Impairments imposed on the optical signal during transmission include extinction ratio degradation, noise accumulation, chirping, and waveform distortion. Re-shaping the optical signal before experiencing signific-ant impairment is effective for extending signal transmission distances [80,81]. Owing to the sinusoidal input–output transfer function [82], the SMZ gate can improve the extinction ratio and compress noise as shown in Figure 3.22. The deviation of the input intensity correspond-ing to symbol '0' from the zero level can be offset and thus the extinction ratio is improved. Because the slope of the transfer function is nearly flat around symbol '0' and symbol '1', the noise distribution in the output is compressed. Re-shaping of not only the amplitude but also the phase is an important issue. The SMZ gate can generate low chirping output intrinsically. Furthermore, the chirping of the input is not transferred to the output, because the nonlinear refractive index change used in the SMZ gate is induced incoherently. These properties enable the design rule for the group velocity dispersion of the transmission link to be simplified.

By combining the SMZ gate with clock recovery components, 3R regeneration can be achieved [83–89]. Generally, the clock recovery process does not transfer high frequency jitter from input data pulses to output clock pulses. Thus, in 3R regeneration, the SMZ gate driven by data pulses with jitter modulates clean clock pulses without jitter. To minimize the effect of the above timing mismatch between the data pulses and the clock pulses, a nearly square switching window provided by the SMZ gate is quite useful. Figure 3.23 shows the setup and the results of loop transmission experiments at 40 Gb/s with an optical 3R regenerator

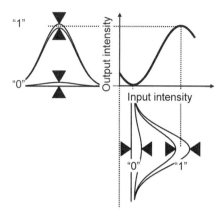

Figure 3.22 Schematic of input–output transfer function

consisting of a cascade of PD-SMZ gates and mode-locked laser diode (MLLD)-based clock recovery [85]. The first PD-SMZ gate is driven by input data pulses and thus modulates clock pulses generated by the first MLLD in a logic-inverted and wavelength-converted manner. These modulated data pulses drive the second PD-SMZ gate and the regenerated output with logic-preserved data at the same wavelength as input. The loop consists of four spans with a total distance of 200 km. Data pulses at 40 Gb/s are regenerated every 200 km. Each span includes a 50-km dispersion-shifted fiber, dispersion compensating fiber, and an EDFA. As shown in Figure 3.23(b), an eye diagram measured at 12 000 km shows clear eye opening. We see no degradation in the receiver sensitivity up to 12 000 km except for the initial step. This is in contrast to the results with transmission without regeneration, which shows large degradation in the receiver sensitivity with increasing transmission distance.

In terms of bit rate transparency, 2R-type signal regeneration is also attractive. In application to 2R regeneration, it has been pointed out that the ultrafast capability of 2R regenerators is necessary because insufficient response speed causes larger timing jitter and thus limits cascadability [90]. The ultrafast capability of the SMZ gate is useful for minimizing the accumulation of timing jitter.

The configuration described above has been focused on the case where the control light is RZ pulses. However, optical communications systems mainly employ the NRZ format at present. The differential phase modulation scheme does not seem immediately applicable to the non-return-to-zero (NRZ) format. However, modifying the operating conditions enables the SMZ gates to be applicable to NRZ operation [91]. Figure 3.24(a) shows the configuration of NRZ wavelength conversion. The structure is the same as the SMZ gate. The gate is driven by injecting NRZ signal light at a wavelength of λ_0 into each SOA. In SOAs, NRZ signal light depletes carriers and thus induces a nonlinear phase shift on CW light at a wavelength of λ_1. To enable NRZ operation, signal light of a different intensity is injected into each arm, because if induced nonlinear phase shifts in both arms are the same in the presence of signal light (corresponding to '1'), the relative phase between both arms remains the same as in the absence of signal light (corresponding to '0'), and thus the data pattern of the wavelength-converted output becomes incorrect. To improve the response speed, the timing of the signal light injected into each arm is displaced.

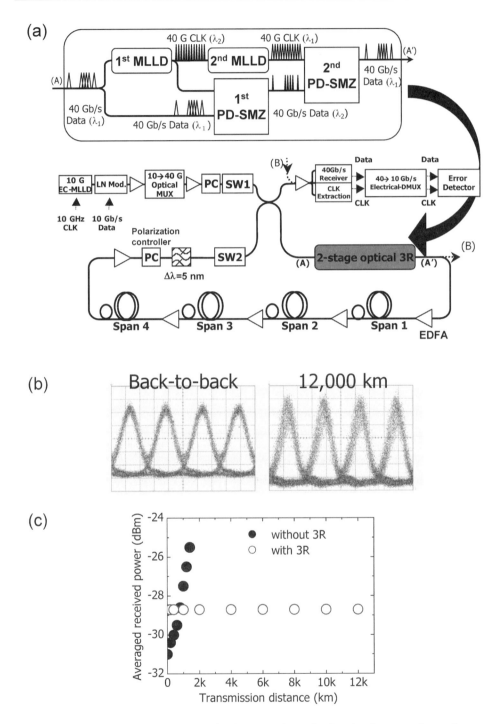

Figure 3.23 Transmission experiments using loop setup for evaluating signal regeneration performance at 40Gb/s: (a) experimental setup; (b) eye diagrams before transmission and after 12 000 km transmission; (c) receiver sensitivity as a function of transmission distance

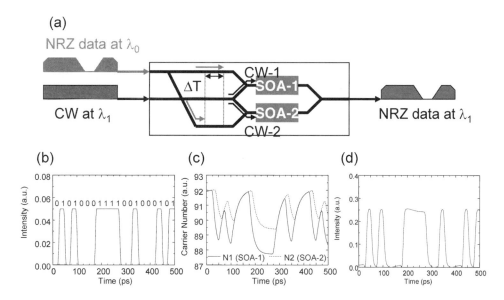

Figure 3.24 Operation scheme for all-optical wavelength conversion of a NRZ signal and simulated results: (a) configuration; (b) input waveform at 40 Gb/s; (c) total carrier numbers in the two SOAs; (d) output waveform at 40 Gb/s

Figures 3.24(b) to (d) show simulated results based on the rate equation model. When NRZ signal light with a waveform shown in Figure 3.24(b) is input, the total carrier number in each SOA (N1 for SOA-1 and N2 for SOA-2) changes as shown in Figure 3.24(c). The data pattern assumed here is '01010001111001000101' at 40 Gb/s. When the input changes from '0' to '1', N1 first quickly decreases but afterward asymptotically approaches a certain value. Also, when the input changes from '1' to '0', N1 first quickly increases but afterward asymptotically approaches another certain value. To take advantage of the fast beginning part of the transition, the excitation of SOA-2 in which lower nonlinear phase shift is induced is delayed. When the delay ΔT is set at 12 ps, the wavelength-converted output shown in Figure 3.24(d) is obtained. Note that the delay ΔT is associated with the carrier lifetime and is independent of the bit rate. Thus, with this configuration, bit-rate-transparent operation can be achieved.

Figures 3.25(a) to (c) show the experimentally measured eye diagrams of NRZ signal wavelength conversion using the HI-SMZ gate, which was also used for RZ signal wavelength conversion. Typical bias current applied to each SOA is also about 300 mA. In this experiment, NRZ signal light at a wavelength of 1558 nm, together with CW light at 1553 nm, is input. The average powers of the NRZ signal light coupled to each SOA are 0 dBm and −10 dBm, respectively. The average power of the CW light coupled to each SOA is about 7 dBm. The delay of signal light into each SOA is set at 12 ps. Under the same operating conditions, wavelength conversion was carried out at 2.5, 10.5, and 42 Gb/s. Figure 3.25(d) shows the results of BER measurement, indicating error-free operation with little power penalty. Measuring BERs as a function of the threshold of the BER tester is a method for evaluating the signal quality (Q-factor), as long as the threshold is adjustable. As shown in Figure 3.25(e), there is little degradation in the Q-factor in operation at 10.5 Gb/s.

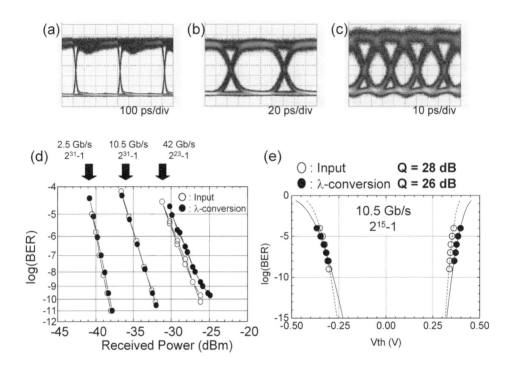

Figure 3.25 Experimental results for NRZ signal wavelength conversion: (a) – (c) wavelength-converted data eye diagrams at 2.5 Gb/s, 10.5 Gb/s, and 42 Gb/s; (d) measured bit error rates; (e) evaluation of Q-factor at 10.5 Gb/s

Applications as signal regenerators are not limited to long-haul transmissions. The property of incoherently regenerating optical signals is useful for all-optical conversion from low-extinction-ratio, high-chirp, short-reach signals generated by a directly modulated laser diode (DM-LD) to high-extinction-ratio, low-chirp, long-reach signals. We experimentally demonstrate that optical signals at 10 Gb/s generated by a DM-LD are converted to higher quality optical signals useful for transmission over an 80-km standard single mode fiber (SMF) without dispersion compensation [92]. Even when the extinction ratio of the input optical signal generated by a DM-LD is low, the deviation of the input intensity corresponding to symbol '0' from the zero level is offset, and thus the extinction ration is improved. The re-shaping property on the phase is another significant feature. The chirp of the input is not transferred to the output, because the nonlinear refractive index change used in the all-optical switch is induced incoherently with carrier density change. The output generated by an interferometric device has low chirp intrinsically. Furthermore, blue-chirp characteristics suitable for transmission through SMF having anomalous chromatic dispersion are expected in the output. The rising edge of the output is accompanied by the decrease in the refractive index change and thus shows a red shift. The falling edge of the output is accompanied by the increase in the refractive index change and thus shows a blue shift. Experiments show that while the output of the DM-LD does not exhibit eye opening after transmission through 80-km SMF without dispersion compensation, the SMZ gate output converted from the signal from the DM-LD does exhibit eye opening.

3.6 Summary

SOA is introduced and brief reviews of optical nonlinear property have been given. Then, two ways of applying the SOA for ultrafast signal processing were presented. In the method where only the ultrafast response component of the response is used, there are many reports as referred in the text. Here, to illustrate the operation principle, we presented our own experiments on 3R operation and wavelength conversion. The SMZ gate device does not use the ultrafast response. Cancellation is made for the slow band-to-band recombination component by differential phase modulation schemes to realize ultrafast gating operations. The method realized many ultrafast signal processing procedures such as DEMUX operation, wavelength conversion, 3R operation applied to 12 000 km long-distance transmission, and processing of wide bit rate range NRZ signal.

The SOA is a highly useful device for all-optical signal processing, not only for ultrafast signals but also for 40 Gb/s or less bit-rates. The important remaining task is to realize more compact hybrid or monolithic integrated devices for versatile applications.

References

[1] S. L. Chung, *Physics of Optoelectronic Devices*, John Wiley & Sons, Inc., New York, 1995.

[2] M. Sugawara, N. Hatori, T. Akiyama, Y. Nakata, and H. Ishikawa, 'Quantum-dot optical amplifiers for high bit-rate signal processing over 40 Gb/s.' *Japan J. Appl. Phys.* **40**, Part2, L488–491 (2001).

[3] K. Morito, M. Ekawa, T. Watanabe and Y. Kotaki, 'High-output-power polarization-insensitive semiconductor optical amplifier,' *J. Lightwave Technol.*, **21**, 176–181 (2003).

[4] L. T. Tiemeijer, P. J. A. Thijs, T. V. Dongen, J. J. M. Binsma, E. J. Jansen, P. I. Kuindersma, G. P. M. Cuijpers, and S. Walczyk, 'High-output-power (+ 15 dBm) unidirectional 1310-nm multiple-quantum-well booster amplifier module,' *IEEE Photon. Technol. Lett.*, **7**, 1519–1521 (1995).

[5] A. Borghesani, N. Fensom, A. Scott, G. Crow, L. Johnston, J. King, L. Rivers, S. Cole, S. Perrin, D. Scrase, G. Bonfrate, A. Ellis, and I. Lealman, 'High saturation output (> 16.5 dBm) and low noise figure (<6 dB) semiconductor optical amplifier for C band operation,' Optical Fiber Communication Conference (OFC 2003), Vol. 86 of *OSA Trends in Optics and Photonics Series* (Optical Society of America, Washington, DC, 2003), paper ThO1, pp. 534–536.

[6] K. Dreyer, C. H. Joyner, J. L. Pleumeekers, C. A. Burrus, A. Dentai, B. I. Miller, S. Shunk, P. Sciortino, S. Chandrasekhar, L. Buhl, F. Storz, and M. Farwell, 'High gain mode-adapted semiconductor optical amplifier with 12.4-dBm saturation output power at 1550 nm,' *J. Lightwave Technol.* **20**, 718–721(2002).

[7] A. E. Kelly, C. Tombling, C. Michie, and I. Andonovic, 'High performance semiconductor optical amplifiers,' Optical Fiber Communication Conference (OFC 2004), Vol. 95 of *OSA Trends in Optics and Photonics Series* (Optical Society of America, Washington, DC, 2004), paper ThS1.

[8] Y. H. Lee, A. Chavez-Pirson, S. W. Koch, H. M. Gibbs, S. H. Park, J. Morhange, A. Jeffery, and N. Payghambarian, 'Room-temperature optical nonlinearities in GaAs,' *Phys. Rev. Lett.*, **57**, 2446–2449 (1986).

[9] J. Mark and J. Mørk, 'Subpicosecond gain dynamics in InGaAsP optical amplifiers: Experiment and theory', *Appl. Phys. Lett.*, **61**, 2281–2283 (1992).

[10] A. Mecozzi and J. Mørk, 'Saturation induced by picosecond pulses in semiconductor optical amplifiers', *J. Opt. Soc. Am. B*, **14**, 761–770 (1997).

[11] H. Chayett, S. Ben Ezra, N. Shachar, S. Tzadok, S. Tsadka, and J. Leuthold, 'Regenerative all-optical wavelength converter based on semiconductor optical amplifier and sharp frequency response,' presented at the Optical Fiber Commun. Conf., Los Angeles, CA, Feb. 2004, Paper ThS2.

[12] M. L. Nielsen, B. Lavigne, and B. Dagens, 'Polarity-preserving SOA-based wavelength conversion at 40 Gb/s using bandpass filtering,' *Electron. Lett.*, **39**, 1334–1335 (2003).

[13] Y. Liu, E. Tangdiongga, Z. Li, S. Zhang, H. de Waardt, G. D. Khoe, and H. J. S. Dorren, '80 Gb/s wavelength conversion using a semiconductor optical amplifier and an optical bandpass filter,' *Electron. Lett.*, **41**, 487–489 (2005).

[14] E. Tangdiongga, Y. Liu, J. H. den Besten, M. van Geemert, T. van Dongen, J. J. M. Binsma, H. de Waardt, G. D. Khoe, M. K. Smit, and H. J. S. Dorren, 'Monolithically integrated 80-Gb/s AWG-based all-optical wavelength converter', *IEEE Photon. Technol. Lett.*, **18**, 1627–1629 (2006).

[15] Y. Liu, E. Tangdiongga, Z. Li, S. Zhang, H. de Waardt, G. D. Khoe, and H. J. S. Dorren, 'Error-free all-optical wavelength conversion at 160 gb/s using a semiconductor optical amplifier and an optical bandpass filter,' *J. Lightwave Technol.*, **24**, 230–236 (2006).

[16] Y. Liu, E. Tangdiongga, Z. Li, H. de Waardt, A. M. J. Koonen, G. D. Khoe, X. Shu, I. Bennion, and H. J. S. Dorren, 'Error-free 320-Gb/s all-optical wavelength conversion using a single semiconductor optical amplifier,' *J. Lightwave Technol.*, **25**, 103–108 (2007).

[17] E. Tangdiongga, Y. Liu, H. de Waardt, G. D. Khoe, and H. J. S. Dorren, '320-to-40-Gb/s demultiplexing using a single SOA assisted by an optical filter', *IEEE Photon. Technol. Lett.*, **18**, 908–910 (2006).

[18] Y. Ueno, S. Nakamura, K. Tajima, and S. Kitamura, '3.8-THz wavelength conversion of picosecond pulses using a semiconductor delayed-interference signal-wavelength converter (DISC)', *IEEE Photon. Technol. Lett.*, **10**, 346–348 (1998).

[19] J. Leuthold, B. Mikkelsen, G. Raybon, C. H. Joyner, J. L. Pleumeekers, B. I. Miller, K. Dreyer, and R. Behringer, 'All-optical wavelength conversion between 10 and 100 Gb/s with SOA delayed-interference configuration', *Opt. Quantum Electron.*, **33**, 939–952 (2001).

[20] M. L. Nielsen and J. Mørk, 'Increasing the modulation bandwidth of semiconductor-optical-amplifier-based switches by using optical filtering', *J. Opt. Soc. Am. B*, **21**, 1606–1619 (2004).

[21] M. L. Nielsen, J. Mørk, R. Suzuki, J. Sakaguchi, and Y. Ueno, 'Experimental and theoretical investigation of the impact of ultra-fast carrier dynamics on high-speed SOA-based all-optical switches', *Opt. Express*, **14**, 331–347 (2005).

[22] J. Mørk and A. Mecozzi, 'Theory of ultrafast optical response of active semiconductor waveguide', *J. Opt. Soc. Am. B*, **13**, 1803–1816 (1996).

[23] G. P. Agrawal and N. A. Olsson, 'Self-phase modulation and spectral broadening of optical pulses in semiconductor laser amplifiers', *IEEE J. Quantum Electron.*, **25**, 2297–2306 (1989).

[24] S. Fu, J. Dong, P. Shum, L. Zhang, X. Zhang, and D. Huang, 'Experimental demonstration of both inverted and non-inverted wavelength conversion based on transient cross phase modulation of SOA', *Opt. Express*, **14**, 7587–7593 (2006).

[25] J. M. Vazquez, Z. Li, Y. Liu, E. Tangdiongga, S. Zhang, D. Lenstra, G. D. Khoe, and H. J. S. Dorren, 'Optimization of optical band-pass filters for all-optical wavelength conversion using genetic algorithms,' *IEEE J. Quantum Electron.*, **43**, 57–63 (2007).

[26] H. Tsuchida and M. Suzuki, '40-Gb/s clock recovery using an injection-locked optoelectronic oscillator', *IEEE Photon. Technol. Lett.*, **17**, 211–213 (2005).

[27] H. Tsuchida, '40-GHz subharmonic optical clock recovery using an injection-locked optoelectronic oscillator', *IEICE Electron. Express*, **3** (15), 373–378 (2006).

[28] H. Tsuchida, '160-Gb/s optical clock recovery using a regeneratively mode-locked laser diode', *IEEE Photon. Technol. Lett.*, **18**, 1687–1689 (2006).

[29] K. Tajima, 'All-optical switch with switch-off time unrestricted by carrier lifetime,' *Jpn. J. Appl. Phys.*, **32**, L1746–L1749 (1993).

[30] K. Tajima, S. Nakamura, and Y. Sugimoto, *Appl. Phys. Lett.*, **67**, 3709–3711 (1995).

[31] N. S. Patel, K. L. Hall, K. A. Rauschenbach '40-Gbits cascadable all-optical logic with an ultrafast nonlinear interferometer,' *Opt. Lett.*, **21**, 1466–1468 (1996).

[32] S. Nakamura, K. Tajima, and Y. Sugimoto, '10 ps all-optical switching in novel Mach–Zehnder configuration based on band-filling nonlinearity of GaAs,' *Conference on Lasers and Electro-Optics* (CELO'94), CThS2, 1994.

[33] J. P. Sokoloff, P. R. Prucnal, I. Glesk, and M. Kane 'A terahertz optical asymmetric de-multiplexer (TOAD),' *IEEE Photon. Technol. Lett.*, **5**, 787–790 (1993).

[34] P. Toliver, R. J Runser, I. Glesk, and P.R. Prucnal 'Comparision of three nonlinear interferometric optical switch geometries', *Opt. Commun.*, **175**, 365–373 (2000).

[35] K. I. Kang, T. G. Chang, I. Glesk, and P. R. Prucnal 'Comparison of Sagnac and Mach–Zehnder ultrafast all-optical interferometric switches based on a semiconductor resonant optical nonlinearity,' *Appl. Opt.*, **35**, 417–426 (1996).

[36] C. Schubert, J. Berger, S. Diez, H. J. Ehrke, R. Ludwig, U. Feiste, C. Schmidt, H. G. Weber, G. Toptchiyski, S. Randel, and K. Petermann, 'Comparison of interferometric all-optical switches for demultiplexing applications in high-speed OTDM systems,' *J. Lightwave Technol.*, **20**, 618–624 (2002).

[37] K. Tajima, S. Nakamura, Y. Ueno, J. Sasaki, T. Sugimoto, T. Kato, T. Shimoda, M. Itoh, H. Hatakeyama, T. Tamanuki, and T. Sasaki, 'Hybrid integrated symmetric Mach–Zehnder all-optical switch and its ultrafast, high extinction switching,' *Electron. Lett.*, **35**, 2030–2031 (1999).

[38] K. Suzuki, T. Shimizu, S. Nakamura, Y. Ueno, M. Takahashi, T. Tamanuki, A. Furukawa, T. Sasaki, and K. Tajima, 'Hybrid integrated symmetric Mach–Zehnder all-optical switch with phase controllers,' *Photonics in Switching* (PS2002), MoB4 (2002).

[39] T. Tamanuki, S. Nakamura, Y. Ueno, K. Tajima, S. Ae, K. Mori, H. Hatakeyama, and T. Sasaki, 'Dependence of nonlinear phase shift on cavity length of SOA for all-optical signal processing,' *Optical Fiber Communication Conference* (OFC2003) ThX4 (2003).

[40] J. Sasaki, H. Hatakeyama, T. Tamanuki, S. Kitamura, M. Yamaguchi, N. Kitamura, T. Shimoda, M. Kitamura, T. Kato, and M. Itoh, 'Hybrid integrated 4 × 4 optical matrix switch using self-aligned semiconductor optical amplifier gate arrays and silica planer lightwave circuit,' *Electron. Lett.*, **34**, 986–987 (1998).

[41] R. Sato, T. Ito, K. Magari, I. Ogawa, Y. Inoue, R. Kasahara, M. Okamoto, Y. Tohmori, and Y. Suzuki, '10-Gb/s low-input-power SOA-PLC hybrid integrated wavelength converter and its 8-slot unit,' *J. Lightwave Technol.*, **22**, 1331–1337 (2004).

[42] G. Maxwell, B. Manning, M. Nield, M. Harlow, C. Ford, M. Clements, S. Lucas, P. Townley, R. McDougall, S. Oliver, R. Cecil, L. Johnston, A. Poustie, R. Webb, I. Lealman, L. Rivers, J. King, S. Perrin, R. Moore, I. Reid, and D. Scrase, 'Very low coupling loss, hybrid-integrated all optical regenerator with passive assembly,' *European Conference on Optical Communication* (ECOC2002), PD3.5 (2002).

[43] I. Armstrong, I. Andonovic, A. E. Kelly, S. Bonthron, J. Bebbington, C. Michie, C. Tombling, S. Fasham, and W. Johnstone, 'Hybridization platform assembly and demonstration of all-optical wavelength conversion at 10 Gb/s,' *J. Lightwave Technol.*, **23**, 1852–1859 (2005).

[44] M. Hattori, K. Nishimura, R. Inohara, and M. Usami, 'Bidirectional data injection operation of hybrid integrated SOA–MZI all-optical wavelength converter,' *J. Lightwave Technol.*, **25**, 512–519 (2007).

[45] M. Schilling, K. Daub, W. Idler, D. Baums, U. Koerner, E. Lach, G. Laube, and K. Wunstel, 'Wavelength converter based on integrated all-active three-port Mach–Zehnder interferometer,' *Electron. Lett.*, **30**, 2128–2130 (1994).

[46] E. Jahn, N. Agrawal, M. Arbert, H.-J. Ehrke, D. Franke, R. Ludwig, W. Pieper, H. G. Weber, and C. M. Weinert, '40 Gbit/s all-optical demultiplexing using a monolithically integrated Mach–Zehnder interferometer with semiconductor laser amplifiers,' *Electron. Lett.*, **31**, 1857–1858 (1995).

[47] D. Wolfson, A. Kloch, T. Fjelde, C. Janz, B. Dagens, and M. Renaud, '40-Gb/s all-optical wavelength conversion, regeneration, and demultiplexing in an SOA-based all-active Mach–Zehnder interferometer,' *IEEE Photon. Technol. Lett.*, **12**, 332–334 (2000).

[48] L. H. Spiekman, U. Koren, M. D. Chien, B. I. Miller, J. M. Wiesenfeld, and J.S. Perino, 'All-optical Mach–Zehnder wavelength converter with monolithically integrated DFB probe source,' *IEEE Photon. Technol. Lett.*, **9**, 1349–1351 (1997).

[49] J. Leuthold, C. H. Joyner, B. Mikkelsen, G. Raybon. J. L. Pleumeekers, B. I. Miller, K. Dreyer and C. A. Burrus, '100 Gbit/s all-optical wavelength interference configuration conversion with integrated SOA delayed-interference configuration,' *Electron. Lett.*, **36**, 1129–1130 (2000).

[50] B. Mikkelsen, K. S. Jepsen, M. Vaa, H. N. Poulsen, K. E. Stubkjaer, R. Hess, M. Duelk, W. Vogt, E. Gamper, E. Gini, P. A. Besse, H. Melchior, S. Bouchoule, and F. Devaux, 'All-optical wavelength converter scheme for high speed RZ signal formats,' *Electron. Lett.*, **33**, 2137–2139 (1997).

[51] M. L. Mašanovic, V. Lal, J. A. Summers, J. S. Barton, E. J. Skogen, L. G. Rau, L. A. Coldren, and D. J. Blumenthal, 'Widely tunable monolithically integrated all-optical wavelength converters in InP,' *J. Lightwave. Technol.*, **23**, 1350–1362 (2005).

[52] X. Song, Z. Zhang, and Y. Nakano, 'Monolithically integrated SOA-MZI all-optical switch with high-yield regrowth-free selective area MOVPE,' *European Conference on Optical Communication* (ECOC2004), Mo3.4.5 (2004).

[53] H. Nakamura, Y. Sugimoto, K. Kanamoto, N. Ikeda, Y. Tanaka, Y. Nakamura, S. Ohkouchi, Y. Watanabe, K. Inoue, H. Ishikawa, and K. Asakawa, 'Ultra-fast photonic crystal/quantum dot all-optical switch for future photonic networks,' *Opt. Express*, **12**, 6606–6614 (2004).

[54] S. Nakamura, Y. Ueno, K. Tajima, J. Sasaki, T. Sugimoto, T. Kato, T. Shimoda, M. Itoh, H. Hatakeyama, T. Tamanuki, and T. Sasaki, 'Demultiplexing of 168-Gb/s data pulses with a hybrid-integrated symmetric Mach–Zehnder all-optical switch,' *IEEE Photon. Technol. Lett.*, **12**, 425–427 (2000).

[55] S. Diez, C. Schubert, R. Ludwig, H.-J. Ehrke, U. Feiste, C. Schmidt, and H. G. Weber, '160 Gbit/s all-optical demultiplexing using hybrid gain-transparent SOA Mach-Zehnder interferometer,' *Electron. Lett.*, **36**, 1484–1486 (2000).

[56] T. Tekin, M. Schlak, W. Brinker, J. Berger, C. Schubert, B. Maul, and R. Molt, 'Ultrafast all-optical demultiplexing performance of monolithically integrated bad gap shifted Mach–Zehnder interferometer,' *European Conference on Optical Communication* (ECOC2001), Th.F.1.3 (2001).

[57] M. Heid, S. Spälter, G. Mohs, A. Färbert, W. Vogt, and H. Melchior, '160-Gbit/s demultiplexing based on a monolithically integrated Mach–Zehnder interferometer,' *European Conference on Optical Communication* (ECOC2001), PD.B.1.8 (2001).

[58] A. Suzuki, X. Wang, T. Hasegawa, Y. Ogawa, S. Arahira, K. Tajima and S. Nakamura, '8 × 160 Gb/s (1.28Tb/s) DWDM/OTDM unrepeatered transmission over 140 km standard fiber by semiconductor-based devices,' *European Conference on Optical Communication* (ECOC2003) (2003).

[59] A. Suzuki, X. Wang, Y. Ogawa and S. Nakamura, '10 × 320 Gb/s (3.2Tb/s) DWDM/OTDM transmission in C-band by semiconductor-based devices,' *European Conference on Optical Communication* (ECOC 2004), Th4.1.7 (2004).

[60] S. Nakamura, Y. Ueno, J. Sasaki, and K. Tajima, 'Error-free demultiplexing at 252 Gbit/s and low-power-penalty, jitter-tolerant demultiplexing at 168 Gbit/s with integrated symmetric Mach–Zehnder all-optical switch,' *European Conference on Optical Communication* (ECOC 2001), Th.F.2.2 (2001)

[61] S. Nakamura, Y. Ueno, and K. Tajima, 'Femtosecond switching with semiconductor-optical-amplifier-based symmetric Mach–Zehnder-type all-optical switch,' *Appl. Phys. Lett.*, **78**, 3929–3931 (2001).

[62] S. Nakamura, Y. Ueno, and K. Tajima, 'Error-free all-optical demultiplexing at 336 Gb/s with a hybrid-integrated symmetric-Mach–Zehnder switch,' *Optical Fiber Communication Conference* (OFC2002), FD3 (2002).

[63] R. J. Manning, A. D. Ellis, A. J. Poustie, and K. J. Blow, 'Semiconductor laser amplifiers for ultrafast all-optical signal processing,' *J. Opt. Soc. Am.*, **B14**, 3204–3216 (1997).

[64] M. Usami, M. Tsurusawa, and Y. Matsushima, 'Mechanism for reducing recovery time of optical nonlinearity in semiconductor laser amplifier,' *Appl. Phys. Lett.*, **72**, 2657–2659 (1998).

[65] T. Durhuus, B. Mikkelsen, C. Joergensen, S. L. Danielsen, and K. E. Stubkjar, 'All optical wavelength conversion by semiconductor optical amplifiers,' *J. Lightwave Technol.*, **14**, 942–954 (1996).

[66] C. Joergensen, S. L. Danielsen, K. E. Stubkjaer, M. Schilling, K. Daub, P. Doussiere, F. Pommerau, P. B. Hansen, H. N. Poulsen, A. Kloch, M. Vaa, B. Mikkelsen, E. Lach, G. Laube, W. Idler, and K. Wunstel, 'All-optical wavelength conversion at bit rates above 10 Gb/s using semiconductor optical amplifiers,' *IEEE J. Sel. Top. Quantum Electron.*, **3**, 1168–1180 (1997).

[67] K. Tajima, S. Nakamura, A. Furukawa, and T. Sasaki, 'Hybrid-integrated symmetric Mach–Zehnder all-optical switches and ultrafast signal processing,' *IEICE Trans. Electron.*, **E87-C**, 1119–1125 (2004).

[68] Y. Ueno, S. Nakamura, and K. Tajima, 'Nonlinear phase shifts induced by semiconductor optical amplifiers with control pulses at repetition frequencies in the 40–160 GHz range for use in ultrahigh-speed all-optical signal processing', *J. Opt. Soc. Am.*, **B19**, 2573–2589 (2002).

[69] M. L. Nielsen, M. Nord, M. N. Petersen, B. Dagens, A. Labrousse, R. Brenot, B. Martin, S. Squedin, and M. Renaud, '40 Gbit/s standard mode wavelength conversion in all-active MZI with very fast response,' *Electron. Lett.*, **39**, 385–386 (2003).

[70] T. Hatta, T. Miyahara, Y. Miyazaki, K. Takagi, K. Matsumoto, T. Aoyagi, K. Motoshima, K. Mishina, A. Maruta, and K. Kitayama, 'Polarization-insensitive monolithic 40-Gbps SOA–MZI wavelength converter with narrow active waveguides,' *IEEE J. Select. Topics Quantum Electron.* **13**, 32–39 (2007).

[71] Y. Miyazaki, K. Takagi, K. Matsumoto, T. Miyahara, T. Hatta, S. Nishikawa, T. Aoyagi, and K. Motoshima, 'Design and fabrication of 40 Gbps-NRZ SOA-MZI all-optical wavelength converters with submicron-width bulk InGaAsP active waveguides,' *IEICE Trans, Electron.*, **E90-C**, 1118–1123 (2007).

[72] L. Billès, J. C. Simon, B. Kowalski, M. Henry, G. Michaud, P. Lamouler, and F. Alard, '20 Gbit/s optical 3R regenerator using SOA based Mach–Zehnder interferometer gate,' *European Conference on Optical Communication* (ECOC97), p. 269 (1997).

[73] B. Mikkelsen, K. S. Jepsen, M. Vaa, H. N. Poulsen, K. E. Stubkjaer, R. Hess, M. Duelk, W. Vogt, E. Gamper, E. Gini, P.A. Besse, H. Melchior, S. Bouchoule and F. Devaux, 'All-optical wavelength converter scheme for high speed RZ signal formats,' *Electron. Lett.*, **33**, 2137–2139 (1997).

[74] A. E. Kelly, I. D. Phillips, R. J. Manning, A. D. Ellis, D. Nesset, D. G. Moodie, and R. Kashyap, '80 Gbit/s all-optical regenerative wavelength conversion using semiconductor optical amplifier based interferometer,' *Electron. Lett.*, **35**, 1477–1478 (1999).

[75] J. Leuthold, C. H. Joyner, B. Mikkelsen, G. Raybon, J. L. Pleumeekers, B. I. Miller, K. Dreyer, and C. A. Burrus, '100 Gbit/s all-optical wavelength conversion with integrated SOA delayed-interference configuration,' *Electron. Lett.*, **36**, 1129–1130 (2000).

[76] Y. Ueno, S. Nakamura, and K. Tajima, 'Penalty-free error-free all-optical data pulse regeneration at 84 Gb/s by using a symmetric-Mach–Zehnder-type semiconductor regenerator,' *IEEE Photonics Technol. Lett.*, **13**, 469–471 (2001).

[77] S. Nakamura, Y. Ueno, and K. Tajima, '168-Gb/s all-optical wavelength conversion with a symmetric-Mach–Zehnder-type switch,' *IEEE Photon. Technol. Lett.*, **13**, 1091–1093 (2001).

[78] S. Nakamura, T. Tamanuki, M. Takahashi, T. Shimizu, S. Ae, K. Mori, A. Furukawa, T. Sasaki, and K. Tajima, 'Ultrafast optical signal processing with hybrid-integrated symmetric Mach–Zehnder all-optical switches,' ITCOM2003 SPIE paper number 5246–42 (2003).

[79] J. Leuthold, L. Möller, J. Jaques, S. Cabot, L. Zhang, P. Bernasconi, M. Cappuzzo, L. Gomez, E. Laskowski, E. Chen, A. Wong-Foy, and A. Griffin, '160 Gbit/s SOA all-optical wavelength converter and assessment of its regenerative properties,' *Electron. Lett.*, **40**, 554–555 (2004).

[80] P. Öhlén and E. Berglind, 'Noise accumulation and BER estimates in concatenated nonlinear optoelectronic repeaters,' *IEEE Photon. Technol. Lett.*, **9**, 1011–1013 (1997).

[81] J. Mørk, F. Öhman, and S. Bischoff, 'Analytical expression for the bit error rate of cascaded all-optical regenerators,' *IEEE Photon. Technol. Lett.*, **15**, 1479–1481 (2003).

[82] B. Mikkelsen, S. L. Danielsen, C. Joergensen, R. J. S. Pedersen, H. N. Poulsen, and K. E. Stubkjaer, 'All-optical noise reduction capability of interferometric wavelength converters,' *Electron. Lett.*, **32**, 566–567 (1996).

[83] R. J. S. Pedersen, M. Nissov, B. Mikkelsen, H. N. Poulsen, K. E. Stubkjaer, M. Gustavsson, W. van Berlo, and M. Janson, 'Transmission through a cascade of 10 all-optical interferometric wavelength converter spans at 10 Gbit/s,' *Electron. Lett.*, **32**, 1034–1035 (1996).

[84] H. J. Thiele, A. D. Ellis, and I. D. Phillips, 'Recirculating loop demonstration of 40 Gbit/s all-optical 3R data regeneration using a semiconductor nonlinear interferometer,' *Electron. Lett.*, **35**, 230–231 (1999).

[85] Y. Hashimoto, R. Kuribayashi, S. Nakamura, K. Tajima, and I. Ogura, 'Transmission at 40 Gbps with a semiconductor-based optical 3R regenerator,' *European Conference on Optical Communication* (ECOC2003), Mo4.3.3 (2003).

[86] R. Inohara, M. Tsurusawa, K. Nishimura, and M. Usami, *European Conference on Optical Communication* (ECOC2003), Mo4.3.2 (2003).

[87] B. Lavigne, P. Guerber, P. Brindel, E. Balmefrezol, and B. Dagens, *European Conference on Optical Communication* (ECOC2001) We.F.2.6 (2001).

[88] O. Leclerc, B. Lavigne, E. Balmefrezol, P. Brindel, L. Pierre, D. Rouvillain, and F. Seguineau, 'Optical regeneration at 40 Gb/s and beyond,' *J. Lightwave Technol.*, **21**, 2779–2790 (2003).

[89] J. Leuthold, R. Ryf, D. N. Maywar, S. Cabot, J. Jaques, and S. S. Patel, 'Nonblocking all-optical cross connect based on regenerative all-optical wavelength converter in a transparent demonstration over 42 nodes and 16 800 km,' *J. Lightwave Technol.*, **21**, 2863–2870 (2003).

[90] P. Öhlén and E. Berglind, 'BER caused by jitter and amplitude noise in limiting optoelectronic repeaters with excess bandwidth,' *IEE Proc.-Optoelectron.*, **145**, 147–150 (1998).

[91] S. Nakamura and K. Tajima, 'Bit-rate-transparent non-return-to-zero all-optical wavelength conversion at up to 42 Gb/s by operating symmetric-Mach–Zehnder switch with new scheme,' *Optical Fiber Communication Conference* (OFC2004), FD3 (2004).

[92] S. Nakamura, 'All-optical conversion from short-reach signal to long-reach signal and its operation in wide wavelength range,' *Optical Fiber Communication Conference* (OFC2005), OTuG4 (2005).

4

Uni-traveling-carrier Photodiode (UTC-PD) and PD-EAM Optical Gate Integrating a UTC-PD and a Traveling Wave Electro-absorption Modulator

Hiroshi Ito and Satoshi Kodama

4.1 Introduction

Among the various semiconductor-based devices for optical signal processing, practical candidates are those using opto-electronic effects, by which an optical signal is converted to an electrical signal (O/E conversion) or vice versa (E/O conversion). Typical examples are laser diodes (LDs), optical signal modulators, and photodiodes (PDs). The operation of these devices is based on relatively simple, well-known physical effects such as photon generation by carrier recombination, modulation of the intensity or phase of light by applying an electric field, or injecting carriers to change the absorption coefficient or refractive index, and carrier generation by photon absorption. These devices have numerous important features: they offer compactness, low power dissipation, high reliability, low cost, ease of design with predictable performance, and they are relatively insensitive to environmental conditions such as temperature, humidity, pressure, and vibration. In addition, they are already widely used in real systems and their fabrication processes and packaging techniques are well established in the industry. By combining these elemental opto-electronic (OE) devices, 'all-optical' signal processing devices can be constructed in, for example, an O/E/O configuration. Figure 4.1(a) schematically shows an elemental configuration of an OE device. Here, a two- or three-terminal device having one output port is considered. Thus, each terminal can be used as a high-speed optical or electrical signal (input or output) port. Although there are 12 possible combinations based

Ultrafast All-Optical Signal Processing Devices   Edited by Hiroshi Ishikawa
© 2008 John Wiley & Sons, Ltd

on the kind of input/output port as shown in Table 4.1, some are not OE device configurations, and two have the same configuration. Therefore, only six can be categorized as independent OE device configurations. Furthermore, there are currently no realistic candidates for configurations 6 and 11, because these configurations are considered to be attainable only as an opto-electronic integrated circuit (OEIC) and thus they are not the elemental devices. Thus, we finally have four variations as elemental devices. Definite examples of each configuration are photodiodes (No. 2), laser diodes (No. 3), electro-absorption modulators (EAMs) and Mach–Zehnder modulators (MZMs) (No. 7), and photo-conductive switches (No. 8). If we try to construct an integrated device by combining two of these elemental devices with up to three input/output ports, we have only three possible configurations as shown in Figure 4.1(b). We can construct configuration (i) by using a combination of elemental devices No. 2 and 3, (ii) by using No. 3 and 6, and (iii) by using No. 2 and 7. Each configuration has optical input/output ports only and thus functions as an all-optical signal processing device. In each case, an electrical signal connects the two elemental devices. In general, the intensity of the electrical signal generated by an O/E conversion device is not high enough to directly drive an E/O conversion device, which means that an electrical amplifier in the electrical signal line is required. However, with current electronics technology, the operating speed of electrical amplifiers is rather insufficient for handling signals beyond 160 Gbit/s, and the additional thermal noise generated by them is also a penalty for obtaining a high sensitivity. These limitations on the performance of electrical amplifiers have restricted the application of O/E, E/O, and O/E/O devices for ultrafast optical signal processing.

The idea for overcoming these limitations is to realize a high-speed and high-output O/E conversion device, since it can eliminate the speed-limiting electrical amplifier, and can open up a new class of opto-electronic integrated device, so that the superior performance of each elemental device is effectively fused as an integrated functional device. This concept

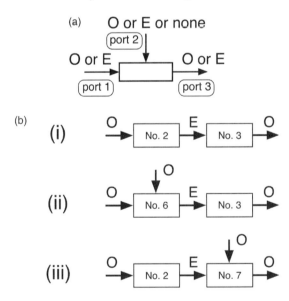

Figure 4.1 Schematic configurations of opto-electronic devices: (a) elemental unit with two or three terminals; (b) possible combinations of two elemental units to construct all optical devices

Table 4.1 Possible configurations and examples of two- or three-terminal opto-electronic devices with one output port, classified by the kind of input/output port

No	input/output port			examples	category
	1	2	3		
1	O	–	O	–	all-optical device
2	O	–	E	PD	OE device
3	E	–	O	LD, MLLD	OE device
4	E	–	E	–	electron device
5	O	O	O	–	all-optical device
6	O	O	E	OEIC?	OE device
7	O	E	O	EAM, MZM	OE device
8	O	E	E	PC switch	OE device
9	E	O	O	–	same as No. 7
10	E	O	E	–	same as No. 8
11	E	E	O	OEIC?	OE device
12	E	E	E	–	electron device

was realized in a novel photodiode, named the uni-traveling-carrier photodiode (UTC-PD) [1] developed by NTT Laboratories in 1996. The UTC-PD can generate high-speed and high-output signal directly to drive conventional E/O devices [2] and is key to realizing ultrafast O/E/O-type all-optical signal processing devices without employing electrical amplifiers. By effectively using this important feature of the UTC-PD, several types of ultrafast opto-electronic integrated device (not only all-optical devices) have been developed [3–7]. Among these, a typical and important example of all-optical signal processing devices is an optical gate integrating a UTC-PD and an EAM with a traveling-wave electrode, which was named as PD-EAM optical gate [4] (corresponding to configuration (iii) in Figure 4.1(b)). The PD-EAM can be used as a multiplexer (MUX), demultiplexer (DEMUX) [8], wavelength converter [9], 2R (reshaping and regenerating) repeater [10], and optical sampling gate [11], and thus can be considered as a fundamental elemental device in optical communications systems.

This chapter covers recent advances in the UTC-PD and PD-EAM optical gate. In the first part of this chapter the basic concept and fundamental performance of the UTC-PD will be discussed in detail. Some results for a high-bit-rate front-end optical receiver, which is another important application of the UTC-PD in ultrafast optical communications systems, will also be presented. The next part of the chapter will cover the basic concept and examples of a new class of opto-electronic integrated device. Finally, the performance of a novel PD-EAM optical gate will be discussed with a view towards future advances in O/E/O-type integrated devices for ultrafast all-optical signal processing.

4.2 Uni-traveling-carrier Photodiode (UTC-PD)

4.2.1 Operation

4.2.1.1 Background

As an optical-to-electrical signal interface device, a PD is most widely used in optical communications systems and optical signal measurement systems. In general, bandwidth and efficiency are important figures of merit for high-speed PDs, though constraints exist due to the tradeoff between these characteristics [12]. In addition, the high-output capability of a PD has also become very important for making full use of the superior features of optical components, such as their wide bandwidth and wide frequency tunability. This is because, the combination of a high-output PD and a broadband optical fiber amplifier (OFA) can eliminate post-amplification electronics and thus extend the bandwidth of the entire system [13]. This new O/E converter configuration is also superior in terms of system simplicity and input sensitivity [14]. However, in a conventional pin-PD, there are inherent tradeoffs between the output current and the other important characteristics mentioned above [12, 15].

To overcome these problems, a novel PD, which simultaneously offers a high 3-dB down bandwidth (f_{3dB}) and a high saturation output current, is required. The superior performance of such a PD will drastically expand its application area, not only as a high-speed photo receiver in communications systems [14, 16] but also as a high output-voltage opto-electronic driver for various high-speed devices. In addition, monolithic integrations of the PD with various elements, such as electron devices [3, 6], optical devices [4, 5, 7], and passive components [17], are also promising, since the effective fusion of the excellent features of each element will lead to novel functions with superior performance that exceeds the limits of conventional electronics and OEIC technologies.

4.2.1.2 Basic Concept

As a solution for the problems associated with conventional pin-PDs, the UTC-PD [1] was developed. The UTC-PD provides a high f_{3dB} and a high saturation output-current simultaneously, owing to its unique operation mode where only electrons are the active carriers traveling through the junction depletion layer [2, 18–20]. The band diagram of the UTC-PD is schematically shown in Figure 4.2(a), and that for the conventional pin-PD is shown in Figure 4.2(b) for comparison. The active part of the UTC-PD consists of a neutral (p-type doped) narrow-gap light absorption layer and an undoped (or a lightly n-type doped) wide-gap (depleted) carrier-collection layer. The photo-generated minority electrons in the neutral absorption layer diffuse (and/or drift) into the depleted collection layer. Here, introducing a quasi-field into the absorption layer by means of the band-gap grading and/or doping grading is effective in reducing electron traveling time in the absorption layer [1, 18]. On the other hand, because the p-type doped absorption layer is quasi-neutral and majority holes exist, additional photo-generated holes respond very fast, i.e., within the dielectric relaxation time, by their collective motion. Therefore, the photoresponse of a UTC-PD is determined only by the high-velocity electron transport in the whole device structure. This is an essential difference from the conventional pin-PD, in which both electrons and holes generated in the depleted absorption layer contribute to the response speed and the low-velocity hole-transport determines the total performance [15]. In addition, in the UTC-PD structure, one can effectively use the velocity overshoot of electrons in the thin depletion layer [21]. This qualitatively different operation mode of the UTC-PD provides several advantages over the conventional pin-PD, such as higher operation speed, higher maximum output current, and lower operation voltage.

In a sense, the UTC-PD structure has several similarities with the base-collector (BC) junction of a bipolar transistor and/or heterojunction bipolar transistor (HBT). Therefore, the guidelines for designing the BC junction of high-performance HBTs can also be adopted for the UTC-PD, such as introduction of a built-in field in the p-type-doped absorption layer by means

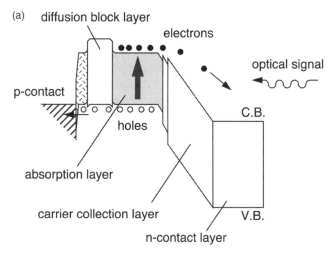

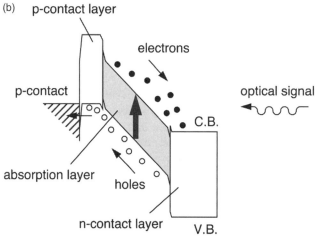

Figure 4.2   Band diagrams of a UTC-PD (a) and a pin-PD (b)

of compositional and/or doping grading [22, 23], and band profile engineering in the collection layer including band-gap grading at the heterointerface [23, 24], doping profile/concentration optimization [25, 26], hybrid (composite) depletion layers [27], and field strength control to make the best use of the high-speed nonequilibrium electron transport in the depletion layer [28, 29].

4.2.1.3 Operation Speed

An advantage of the UTC-PD is its higher device operation speed owing to an order of magnitude higher electron velocity in the depletion layer as compared to the hole velocity. In an appropriate UTC-PD structure with a moderate absorption layer thickness (W_A), electron diffusion time in the absorption layer mainly determines the operation speed. In general, diffusion velocity is considered to be smaller than the drift velocity. However, the electron diffusion velocity in the InGaAs absorption layer can be very large for a relatively thin absorption layer due to the uniquely large minority mobility of electrons in p-InGaAs [30], even if the absorption layer has a uniform band-gap and doping structure. Of course, the conventional techniques mentioned in the previous sub-section (introduction of an electric field [1, 18]) are also effective for enhancing the electron velocity in the absorption layer, and were applied to the UTC-PD structure [31]. In addition, we can independently design the depletion layer and the absorption layer thicknesses in the UTC-PD structure. Thus, a very thin absorption layer can be used to attain an extremely high f_{3dB} without sacrificing the CR charging time. This is also an important advantage of the UTC-PD over the pin-PD, in which the CR charging time becomes significantly larger when the absorption layer thickness is excessively reduced to decrease the carrier transit time [12].

Figure 4.3 compares calculated f_{3dB} for the UTC-PD and pin-PD as a function of W_A. Here, the load resistance of 50 Ω, a constant stray capacitance (the same as the junction capacitance), and a collection layer thickness (W_C) for the UTC-PD of 300 nm are assumed. If we try to increase the f_{3dB} of a pin-PD, we have to decrease W_A. However, as W_A is decreased, the bandwidth becomes smaller again at a certain point due to the junction capacitance increase. In Figure 4.3, calculations for three pin-PDs with different areas (solid curves) are shown. Although the maximum f_{3dB} becomes larger when the device area (S) is smaller, the f_{3dB} of each device peaks at certain W_A and becomes smaller again. The smallest device area in this figure ($S = 10\,\mu m^2$) corresponds to an absorption area diameter of about 3.5 μm. Thus, by considering the practical spot-size limitation of a conventional optical lens system, there is not much room for further improvements of the f_{3dB} for pin-PDs with this approach. On the other hand, in the UTC-PD structure, the absorption layer and the collection layer are separated. Thus, f_{3dB} increases monotonically up to very thin W_A without a CR charging time increase (broken curve in Figure 4.3). Therefore, the predicted maximum f_{3dB} is much larger than that for the pin-PD.

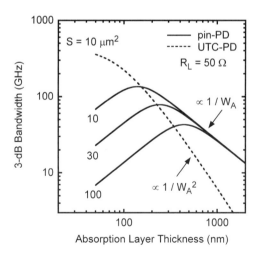

Figure 4.3 Calculated tendencies of 3-dB bandwidth against absorption layer thickness for pin-PDs with different device areas and a UTC-PD

4.2.1.4 Output Current

Another important advantage of the UTC-PD is its higher output saturation current due to the much smaller space charge effect in the depletion layer, which results from the high electron velocity in the depletion layer. In the conventional pin-PD, the band profile is modified under a high optical input condition because photo-generated carriers (holes) are stored and act as space charge in the depleted absorption layer, especially at the n-type contact layer side [32] as shown in Figure 4.4(b). This is because the high-velocity electrons [33] rapidly travel through and are swept out from the depletion layer even with a small electric field. The decreased electric field drastically reduces the hole velocity, enhances the carrier storage (mostly holes), and results in significant output current saturation. Here, the electric field required to maintain saturated hole velocity in the InGaAs depletion layer is very large [34], implying that the output current linearity is easily affected by the electric field reduction due to the space charge accumulation. In addition, when the optical input is turned off, the stored holes can only be swept out gradually because the hole drift velocity increases very slowly with increasing electric field [34]. This situation will result in long output current tailing for a fast optical input signal. Although the situation of the space charge accumulation is similar in the UTC-PD as shown in Figure 4.4(a),

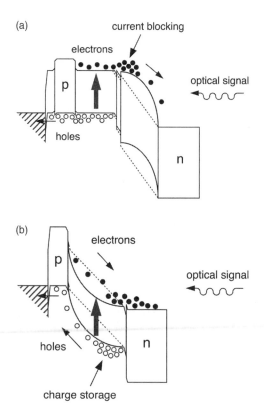

Figure 4.4   Modified band diagrams of (a) UTC-PD, and (b) pin-PD at high optical input

the space charge consists only of electrons whose velocity at overshoot [21,35] is much higher than that of holes, even for the decreased electric field. Therefore, the output current does not saturate until the current density becomes significantly higher than that for the pin-PD. This also means that the linearity range of the UTC-PD is much wider. In addition, different from the pin-PD, the space charge influences the output current by blocking the diffusion current at the absorption/collection layer interface, which is exactly the same phenomenon as the well-known Kirk effect in a bipolar transistor [36]. Thus, once the electric field is recovered, the space charge will be quickly swept out from the depletion layer, resulting in a short fall-off time of the output response.

Another important feature is that the fast response with a high saturation current of the UTC-PD is maintained at a low (or even at zero) bias voltage, because the high electron velocity in the depletion layer can be maintained at a relatively low electric field or even with the built-in field of the InGaAs/InP pn junction (25 kV on average for $W_A = 300$ nm, for example). This makes the high-speed and high-output current operation of the UTC-PD possible without applying any bias voltage. A smaller operation voltage is advantageous in many ways: it reduces power consumption, simplifies heat sinking, eliminates biasing circuit, and improves reliability.

4.2.2 Fabrication and Characterization

4.2.2.1 Epitaxial Layer Structure

A typical epitaxial layer structure for the InP/InGaAs UTC-PD is shown in Figure 4.5. The epitaxial layers are grown on a (100) oriented semi-insulating (S.I.) InP substrate by low-pressure MOCVD. The p-type and n-type dopants are C (or Zn) and Si, respectively. The absorption layer consists of p-InGaAs, thin p^+-InGaAs, and thin undoped InGaAs, and the collection layer consists of thin undoped InGaAsP, thin undoped InP, thin n^+-InP, and lightly n-doped or undoped InP. Here, the p-InGaAs absorption layer is moderately doped (typically from 1 to 20×10^{17}/cm^3) to obtain the benefit from the self-induced field (self-bias effect) in the absorption layer [19,37]. In order to suppress current blocking at the absorption/collection

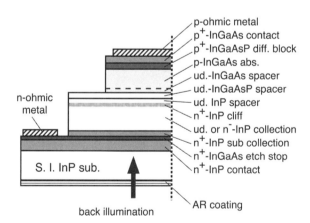

Figure 4.5 Schematic view of a typical UTC-PD epitaxial layer structure with a double-mesa configuration

layer interface, a step-graded band-gap profile is used by inserting an InGaAsP (typical band-gap energy of about 1 eV) layer between the InGaAs absorption and InP collection layers. On the p-type contact layer side, a p^+-InGaAsP carrier diffusion block layer is inserted between the absorption layer and p^+-InGaAs contact layer to prevent the surface recombination of photo-generated electrons. On the n-type contact layer side, an n^+-InGaAs selective-etch stop layer is inserted between n^+-InP sub-collection and contact layers for ease of device fabrication.

4.2.2.2 Fabrication Process and Characterization

To characterize the high-speed device performance, hexagonally or octagonally shaped double-mesa structure devices were fabricated using wet chemical selective etching and metal lift-off processes. Each device was then integrated with a 50-Ω coplanar waveguide (CPW) on the S.I. InP substrate. Basically, the PD output electrode on the p-type contact layer is connected to the signal line, while the n-type contact layer is connected to the ground in the case of photo receiver application, so as to output a negative voltage signal, which is suitable for directly driving the standard digital circuits. Depending on the device configurations, the backside (for back-illuminated devices) or the side (for waveguide or refracting-facet devices) of the substrate was mirror polished (or spontaneously etch-stopped at (111)A facet for refracting-facet devices) and anti-reflection (AR) coated after the devices were fabricated. Figure 4.6 shows an SEM (scanning electron microscope) image of a back-illuminated UTC-PD. The p-type contact electrode is connected to the center electrode of the CPW, and the n-type contact electrode is connected either directly to the ground electrode or to the DC biasing pad through a capacitor located below the ground electrode.

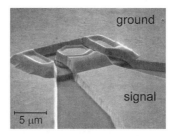

Figure 4.6   SEM micrograph of a typical UTC-PD with a CPW output

The responsivity was typically measured in a broad area device at $\lambda = 1.55\,\mu$m using a continuous wave (CW) light source. For the high-frequency response measurement, two methods were alternatively used. One is a small-signal frequency response measurement using a cascade probe and a lightwave component analyzer (HP-83467C) up to 50 GHz. A CW light ($\lambda = 1.55\,\mu$m) was combined with the small-signal light from the analyzer in order to evaluate the response at higher average photocurrents. The other one is a pulse photoresponse measurement by pump-probe electro-optic sampling (EOS) [38] using a 1.55-μm incident pulse (a full width at a half maximum (FWHM) of 280 or 400 fs, a repetition rate of 30 or 100 MHz, and an external CdTe probe chip. A schematic drawing of the experimental arrangement for the EOS is shown in Figure 4.7. The device under test (DUT) is typically connected to two 50-Ω CPWs (resulting in a 25-Ω load) and biased from one of the CPWs. The other CPW is terminated by a 50-Ω load to eliminate signal reflection. The signal pulse is irradiated form the backside of the device, and the probe pulse is irradiated from the front side onto the CdTe probe chip placed on the CPW adjacent to the DUT. The Fourier transform of the measured pulse response enables us to obtain the relative response of the device at frequencies far exceeding those in conventional network analysis. Another advantage of the EOS is its low average optical input power, which enables us to measure the intrinsic characteristics of the device at very high current densities without seriously suffering from thermal effects.

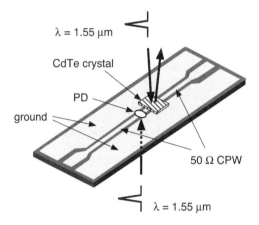

Figure 4.7 Experimental arrangement for EOS measurement

4.2.3 Characteristics of the UTC-PD

4.2.3.1 Basic Photoresponse

First, pulse photoresponses of a UTC-PD and pin-PD are compared to clarify the essential difference in the carrier dynamics of each device. Figures 4.8(a) and (b) show output waveforms of the pulse photoresponses of a UTC-PD and pin-PD [15] measured using the EOS technique. For this measurement, the input optical-pulse energy was changed from 0.2 to 2.0 pJ/pulse and the bias voltage was fixed at -2 V. In order to make a reasonable comparison, both devices were designed to have the same junction capacitance, the same load resistance (25 Ω), and similar f_{3dB} at low input optical energy. The depletion layer thickness was 300 nm and the absorption area was 20 μm^2 for both devices.

In UTC-PD, the output peak current increases linearly with increasing input energy, and the waveform does not significantly change until it reaches the saturation point. More precisely, the pulse width (FWHM) decreases slightly with increasing input energy. This phenomenon will be discussed in a later section. After the saturation occurs, the pulse width increases gradually. However, even at this stage, the fall time of the waveform does not obviously increase. As mentioned in the previous section, the output current saturation is caused by the current blocking due to the space–charge induced modification (reduction) of the electric field in the collection layer. The fast fall time, even in the saturation region, is attributed to the fast

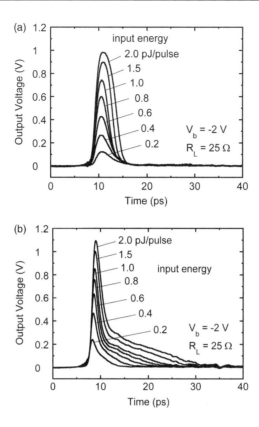

Figure 4.8 Pulse photoresponses of (a) UTC-PD and (b) pin-PD [15]. (Reproduced from by permission of IOP Publishing Limited)

response of electrons accumulated in the depletion layer, which simultaneously implies that, at this fall-off stage, the operation of the device quickly returns to the normal linear response mode. This indicates that the linearity of the output current against input optical power in the UTC-PD is very high, which is an important feature in attaining better signal-to-noise characteristics for both digital and analog applications.

In contrast, the waveform of the pin-PD is quite different, consisting of two characteristic current components. The initial sharp component (FWHM of about 1 ps) is attributed to fast electron transport and the slow tail is caused by hole transport. The peak current of the electron response starts to saturate at relatively low input energy, and the pulse width increases gradually at the same time. More precisely, the fall time of the pulse becomes larger when the input energy becomes larger, while the rise time stays almost constant. These are attributed to the electron velocity decrease caused by the electric field decrease (the space charge effect) in the depletion layer, especially on the n-type contact layer side. On the other hand, the FWHM of the hole-response increases drastically as the input energy increases. This is due to the stronger space charge effect of holes in the depletion (absorption) layer of the pin-PD. The much slower hole response compared with that of electrons produces hole accumulation and electric field reduction at the n-type contact layer side in the depletion layer. The reduction causes even slower hole transport, and this positive feedback results in a significant broadening of the photoresponse waveform, even at a relatively low input energy. The broadening of the output waveform for pin-PD reflects an additional drawback of the pin-PD. Since the holes are accumulated and the electric field is much decreased on the n-type contact layer side, photo-generated electrons are also stored in the depletion layer to maintain the charge neutrality at the n-type contact layer side, making the charge neutral zone wider [32]. Thus, the photo-generated holes can only be swept out very slowly from this accumulation region, resulting in the very long tailing of the output waveform in the pin-PD. The significant tailing of the pulse response results in severe degradation of the bandwidth and linearity of the pin-PD at increased input optical power, which occurs at much smaller input than that causes the output current saturation in the UTC-PD.

This comparison clearly indicates that the UTC-PD is superior for obtaining wide linearity and a very high output current level while maintaining a very fast response.

4.2.3.2 Bandwidth

As mentioned in the previous section, the UTC-PD is advantageous for achieving extremely high speed performance. To demonstrate this feature, an ultimately scaled-down device was fabricated and characterized. Figure 4.9 shows the pulse photoresponse of a back-illuminated InP/InGaAs UTC-PD with a fast-response design in which W_A is 30 nm, W_C is 230 nm, absorption area (S) is 5 μm^2, and load resistance is 12.5 Ω [39]. The narrowest FWHM of 0.97 ps was obtained at a very low bias voltage of -0.5 V. The Fourier transform of this pulse response (after the deconvolution for the incident pulse widening) gives an f_{3dB} of 310 GHz as shown in Figure 4.10 [39]. This is the highest f_{3dB} ever reported for PDs operating at 1.55 μm. This device also exhibited a 10-dB down bandwidth of 750 GHz and a 15-dB down bandwidth of over 1 THz. It is also worth mentioning that the output peak current is as high as 8 mA even with such an ultrafast response.

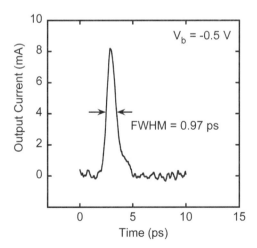

Figure 4.9 Pulse photoresponse of a UTC-PD with a fast-response design [39]. (Reproduced by permission of © 2000 IET)

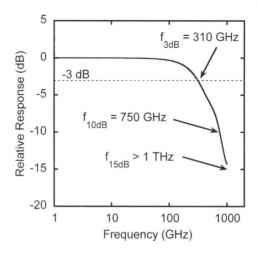

Figure 4.10 Fourier transform of the waveform in Figure 4.9 [39]. (Reproduced by permission of © 2000 IET)

Figure 4.11 summarizes the experimentally obtained relationship between f_{3dB} and W_A for back-illuminated InP/InGaAs UTC-PDs [19, 39]. These data were also obtained from the Fourier transform of pulse photoresponses. Data for conventional pin-PDs are also shown for comparison. The f_{3dB} is basically proportional to $1/W_A^2$ for the UTC-PD and to $1/W_A$ for the pin-PD. This difference comes from the difference in the carrier transport in the absorption layer; namely, the electron transport in the neutral absorption layer is diffusive in the UTC-PD, which is in contrast to the drift motion of both carriers in the depletion layer of the pin-PD. The f_{3dB} at high optical inputs is significantly larger than that at low optical inputs, due to the effect of the self-induced field in the absorption layer [19, 37], which is discussed in detail in the following section.

A simplified expression of the 3-dB bandwidth for the pin-PD [40] is given by:

$$f_{3dB} \cong 3.5/2\pi\,\tau_{ave} \tag{4.1}$$

where τ_{ave} is average carrier traveling time in the depletion layer $(= W_A/v_{ave})$ and v_{ave} the average carrier velocity [40] (which is dominated by the hole velocity). Thus, f_{3dB} is basically inversely proportional to W_A. On the other hand, when the electron transport in the collection layer is dominant, as in the extreme case $(W_A = 0)$, the 3-dB bandwidth for the UTC-PD is similarly expressed [19] as:

$$f_{3dB} \cong 2.8/2\pi\,\tau_e \tag{4.2}$$

where τ_e is average electron traveling time in the collection layer $(= W_C/v_e)$ and v_e the average electron velocity. Note that τ_e is much smaller than τ_{ave} because the electron velocity at overshoot in InP (about 4×10^7 cm/s) [21,35] is much larger than hole drift velocity in InGaAs $(4.8 \times 10^6$ cm/s) [41].

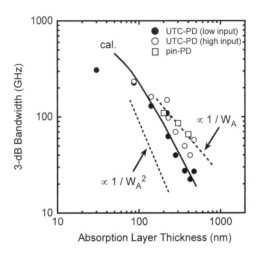

Figure 4.11 Relationships between 3-dB bandwidth and absorption layer thickness for UTC-PDs and pin-PDs [19]. (Reproduced by permission of © 2000 IEICE)

In a more realistic structure with an appropriate W_A, the electron diffusion time in the absorption layer dominates the total carrier transit time in the UTC-PD. In such a case, the 3-dB bandwidth is expressed [19] as:

$$f_{3dB} \cong 1/2\pi(W_A^2/3D_e + W_A/v_{th}) \tag{4.3}$$

where $D_e (= \mu_e kT/q)$ is the diffusion coefficient of minority electrons, v_{th} the thermionic emission velocity of electrons (2.5×10^7 cm/s in InGaAs), μ_e is the minority electron mobility, k the Boltzmann constant, T the temperature, and q the electron charge. This indicates that the f_{3dB} of the UTC-PD is basically inversely proportional to W_A^2 except when the absorption layer is very thin.

The relative frequency response of a photodiode ($B(f)$) is generally expressed as:

$$B(f) = 1/((1 + (f/BW_{CR})^2)(1 + (f/BW_{TR})^2)) \tag{4.4}$$

where f is frequency, BW_{CR} the CR time constant limited bandwidth ($= 1/2\pi CR$), and BW_{TR} the carrier transit time limited bandwidth. Therefore, to discuss the intrinsic carrier transport in a PD, we have to be careful to ensure that the influence of BW_{CR} is relatively small.

The solid curve in Figure 4.11 shows the calculated tendency for the UTC-PD, which considers the contributions of the carrier transit times in the absorption and collection layers [19]. The calculation agrees well with the experimental results at low optical inputs except for the highest f_{3dB} case, where the absorption layer is extremely thin, i.e., 30 nm. Although this discrepancy can be attributed to several factors, such as increased influence of parasitic elements in the device and signal broadening in the electrical waveguide, we should carefully consider the resolution limit of the experimental set-up, such as the finite influence of the pulse width widening caused by the transmission characteristics of optical fibers, the light round-trip time in the CdTe chip, and the finite dimension of the optical probe beam on the waveguide, all of which were not taken into account in this experiment. Thus, further improvement of the evaluation technique is also required to accurately evaluate such ultrafast devices.

4.2.3.3 Carrier Transport in the UTC-PD

One of the advantages of the UTC-PD is that it can take advantage of the electron velocity overshoot to boost operation speed and output current. This phenomenon has also been experimentally studied. Figure 4.12 shows the relationship between the 1/FWHM of pulse photoresponse [39] and the bias voltage (V_b) at an optical input power of 0.14 pJ/pulse in a device with a very thin absorption layer of 30 nm. With increasing negative bias voltage, the 1/FWHM increases first, showing a maximum at a bias voltage of about −0.5 V, and then decreases monotonically with increasing V_b. The V_b of −0.5 V corresponds to an average electric field in the carrier traveling region of about 50 kV/cm, assuming a constant electric field throughout the device and neglecting the operation voltage shift due to the load and stray resistances. Because the absorption layer of the fabricated device is very thin, the contribution of the electron transport in the depletion layer to the total delay time is dominant. Therefore, the decrease in 1/FWHM is mainly attributed to the reduction in the average electron velocity in the depletion region.

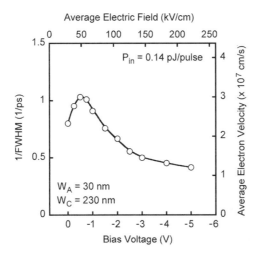

Figure 4.12   Relationship between 1/FWHM of pulse photoresponses and bias voltage

At a bias voltage of −5 V, the output pulse tended to show a trapezoid shape, which is usually seen in time-of-flight experiments. Therefore, from its FWHM of 2.4 ps, we calculated the average electron velocity at $V_b = -5$ V to be 1.1×10^7 cm/s, assuming that electron velocity is constant throughout the device. Thus, considering the influence of the finite width of the pump and probe pulses, we estimated the average electron velocity at the smallest FWHM (0.97 ps) to be 3.0×10^7 cm/s [39]. Here, the influence of the CR charging time on the FWHM was neglected. The converted scale of the average electron velocity is shown on the right vertical axis in Figure 4.12. Although the above assumptions include underestimations, this average electron velocity is about three times larger than the reported electron drift velocity of about 1×10^7 cm/s in undoped InP at a similar electric field [42]. This indicates that the electron velocity overshoot is actually taking place in the depletion layer of the UTC-PD. Similar phenomena have also been observed in the collector layer of HBTs [29, 43, 44]. It is reported that the electron saturation velocity in a long sample peaks at an electric field of about 10 kV/cm in InP [45] due to the inter-valley scattering of electrons with large kinetic energies. However, in the case of a short traveling distance, such as in the depletion layer of the UTC-PD, the optimum electric field providing the shortest average electron traveling time becomes higher due to the nonequilibrium transport of electrons (electron velocity overshoot) [21, 33, 46]. For example, the optimum electric field to minimize the electron traveling time in the 230-nm GaAs layer is calculated to be about 20 kV/cm [21, 30], while the electric field that gives the peak electron velocity in a long sample is about 4 kV/cm [47]. A similar tendency is also seen in InGaAs (about 20 kV/cm [33] versus 3 kV/cm [48]). Therefore, the fact that the optimum electric field that gives the smallest FWHM in the thin InP collection layer of the UTC-PD shown in Figure 4.12 (about 50 kV/cm) is several times larger than that in a long sample (about 10 kV/cm [45]), coincides well with the understanding that there is a considerable contribution by electron velocity overshoot to the electron transport in the depletion layer consisting of InP. These results indicate that, in order to attain ultrahigh-speed operation (as well as very high output current) of a UTC-PD, it is essential to take full advantage of the electron velocity overshoot in the collection layer.

4.2.3.4 Zero-biased Operation

As explained in the previous sections, the UTC-PD is advantageous in terms of its low operation voltage. An extreme case for this is an operation of the UTC-PD without any bias voltage. Eliminating the electric power supply is practically important for technologies such as those for portable measurement instruments, because it makes systems simpler, smaller, lighter, and less expensive. Eliminating conductive cables [49] is also advantageous because it makes systems immune from electrical surge and ensures that the electromagnetic field distribution is not disturbed. For such a requirement, UTC-PD is also promising, since the build-in field in the pn junction depletion region is sufficient for maintaining its high-speed and high-output capabilities. As an example, the ultrafast pulse photoresponse of a UTC-PD with W_A of 30 nm, W_C of 230 nm, and S of 5 μm^2 at zero bias conditions has been demonstrated [50]. Here, the average electric field in the depletion layer is calculated to be about 33 kV/cm, which is sufficient for realizing the electron velocity overshoot in the depletion layer. Despite the fact that no bias voltage was applied, the UTC-PD exhibited a very short pulse response with an FWHM of 1.22 ps. The Fourier transform resulted in an f_{3dB} of 230 GHz. The output peak current was as high as about 7 mA, which is comparable to the typical output peak current in conventional 50-GHz class, high-speed pin-PDs operated at a higher bias voltage, while the obtained f_{3dB} is much higher than the record f_{3dB} of conventional pin-PDs (145 GHz) [51]. A maximum RF output power of 0.58 mW at 100 GHz [52] has also been obtained from a UTC-PD without applying bias voltage.

4.2.3.5 Effect of Self-induced Field

The enhancement of f_{3dB} at high optical inputs shown in Figs. 4.8(a) and 4.11 is one of the unique features of the UTC-PD, and is explained by the effect of the self-induced field in the absorption layer [19, 37]. When the intensity of the input optical signal is relatively high, the concentration of photo-generated holes against that of the majority holes supplied from the dopants in the absorption layer becomes prominent. As a result, the hole drift current makes a certain contribution to the total current flow because the hole current must flow through the absorption layer to maintain the current continuity, such that:

$$J(x)_h + J(x)_e = const. \tag{4.5}$$

where $J(x)_h$ is hole current density at position x and $J(x)_e$ the electron current density. According to the constitutive relation, this hole drift current induces an electric field in the absorption layer, $E(x)_{ind}$, as shown in Figure 4.13 without any external biasing signal, such that:

$$E(x)_{ind} \cong J(x)_h/\sigma_p \cong J(x)_h/qp_0\mu_h \tag{4.6}$$

where σ_p is the conductance of the absorption layer, q the electron charge, p_0 the majority hole concentration of the absorption layer, and μ_h the majority hole mobility in the absorption layer. This automatically induced (self-induced) field in the absorption layer accelerates the photo-generated minority electrons in it. Thus, when the operation current is above a certain level, the electron transport is regarded as drift/diffusion rather than simple diffusion. Such a phenomenon enhances the f_{3dB} of the UTC-PD in a high-excitation condition. In principle, $J(x)_h$ is zero at the absorption/collection layer interface ($x = W_A$), and linearly increases (under the assumption of uniform carrier generation) toward the p-contact layer side ($x = 0$) up to the diode operation photocurrent density, J_{op}. Thus, the maximum induced field, $E(0)_{ind} \cong J_{op}/qp_0\mu_h$, does not depend on the absorption layer thickness, while the diffusion velocity decreases in reverse proportion to W_A^2. This implies that the benefit of this self-induced field is more prominent when the absorption layer is relatively thick. In addition, at the optical signal input, the distribution of photo-generated electron concentration changes dynamically and nonuniformly because of the unidirectional diffusion of electrons to the collection layer, and thus the electron diffusion velocity becomes maximum at $x = W_A$. Therefore, the self-induced field complimentary enhances the electron velocity in the absorption layer, especially at around $x = 0$, to make the average electron transit time in the absorption layer much smaller.

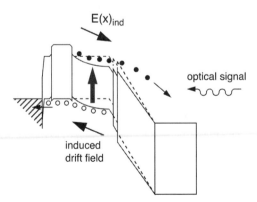

Figure 4.13 Modified band diagram of a UTC-PD when the electric field is spontaneously induced in the absorption layer at high optical input

The total potential drop in the absorption layer, $\Delta\phi_{ind}$, is expressed as:

$$\Delta\phi_{ind} \cong J_{op} W_A / 2q p_0 \mu_h \tag{4.7}$$

This equation indicates that the benefit of the self-induced field is stronger for smaller p_0 and proportional to the photocurrent, $I_p (= J_{op} \times S)$. For example, when $p_0 = 4 \times 10^{17}/cm^3$, $\mu_h = 130\,cm^2/Vs$ [53], and $J_{op} = 100\,kA/cm^2$ ($I_p = 50\,mA$ for $S = 50\,\mu m^2$), $E(0)_{ind}$ is calculated to be $12\,kV/cm$, which results in a drift velocity of $7.2 \times 10^7\,cm/s$ assuming a μ_e of $6000\,cm^2/Vs$ [54]. This is much higher than the pure diffusion velocity of $1.6 \times 10^7\,cm/s$ for $W_A = 300\,nm$, for example, which can well explain the observed significant enhancement of f_{3dB} in Figure 4.11. The potential drop, $\Delta\phi_{ind}$, for the above-mentioned case is calculated to be $180\,mV$, which is smaller than the energy separation between Γ- and L-valley minima for electrons in InGaAs ($550\,mV$ [55]), resulting in the high electron drift velocity in the absorption layer without the inter-valley scattering.

Figure 4.14 shows an example of the dependence of the 2-dB down bandwidth (f_{2dB}) on the average photocurrent (I_p) for a UTC-PD at a V_b of $-3\,V$. Here, the device has a 470-nm thick p-InGaAs absorption layer, with graded doping from $2 \times 10^{18}/cm^3$ to $1 \times 10^{17}/cm^3$, and a 290-nm thick undoped InP collection layer [56]. The absorption area was $99\,\mu m^2$. The f_{2dB} values were estimated from the small-signal frequency responses measured using the lightwave component analyzer. Here, f_{2dB} was employed instead of f_{3dB} because of the frequency range limitation of the analyzer ($50\,GHz$). The f_{2dB} increased rapidly with increasing photocurrent (optical input power) and reached a maximum value. Then it stayed almost constant and finally decreased gradually. The first low f_{2dB} in the low photocurrent region is attributed to the less effective self-induced field in the absorption layer at low excitation, while the decreased f_{2dB} in the high photocurrent region is mainly due to the widening of the neutral region caused by the space charge effect in the collection layer (Figure 4.4(a)). The maximum electric field in the 470-nm thick doping-graded absorption layer is calculated to be $4.2\,kV/cm$ when the photocurrent is $10\,mA$. Even with such a relatively small electric field, the electron drift velocity can

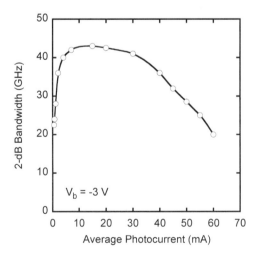

Figure 4.14   Dependence of 2-dB bandwidth of a UTC-PD on average photocurrent

reach about 1.7×10^7 cm/s (assuming a μ_e of 4000 cm²/Vs for InGaAs with $p_0 = 2 \times 10^{18}/cm^3$ [54]), which is considerably larger than the pure diffusion velocity of 7.1×10^6 cm/s, resulting in the drastic increase in f_{2dB} in the low photocurrent region as seen in Figure 4.14.

4.2.3.6 Output Current Saturation

As explained in the previous section, the fact that only electrons travel through the depletion layer provides the high output current of the UTC-PD, which is an important benefit over the conventional pin-PD. Figure 4.15 shows an experimental result of the output peak voltage, measured using the EOS technique, against input optical pulse energy for several bias voltages [19]. Here, the device has the following parameters: $W_A = 220$ nm, $W_C = 300$ nm, $S = 52 \,\mu m^2$, and $R_L = 25 \,\Omega$. As seen in Figure 4.15, the output voltage (output current) increases linearly with increasing input optical pulse energy (optical power), and starts to saturate at certain point depending on the bias voltage. A characteristic feature of the UTC-PD shown here is that the output current (or current density) at which the experimental value starts to deviate from the linear dependence on the input optical energy (more than 60 mA or 115 kA/cm² for $V_b = -3$ V) is much higher than that for conventional high-speed pin-PDs.

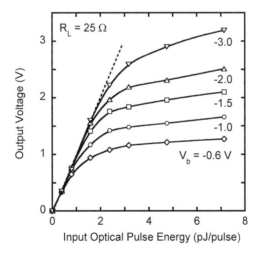

Figure 4.15 Dependence of output peak voltage from a UTC-PD on input optical pulse energy for several bias voltages [19]. (Reproduced by permission of © 2000 IEICE)

If we assume a simple velocity saturation model in which the carrier saturation velocity (v^s) stays constant when the electric field is larger than a certain value (critical electric filed, E^c) and becomes smaller when the electric filed is smaller than E^c, the critical (space charge) densities for electrons in the UTC-PD and for holes in the pin-PD are calculated by solving Poisson's equation as follows [19]:

$$n_c - N_D = 2\varepsilon(-V_b + V_{bi} - E_e{}^c W_C)/qW_C{}^2 \tag{4.8}$$

$$p_c + N_D = 6\varepsilon(-V_b + V_{bi} - E_h{}^c W_A)/qW_A{}^2 \tag{4.9}$$

where n_c is the critical electron density for the UTC-PD, N_D the background donor density in the depletion layer, ε the dielectric constant, V_{bi} the built-in voltage of the pn junction, E_e^c the critical electric field for electrons, p_c the critical hole density for the pin-PD, and E_h^c the critical electric field for holes. Here, (as is normally the case for the actual PD structure) only the n-type residual impurities (donors) in the depletion region are considered and the influence of electrons in the pin-PD is neglected. Thus, the output current saturation starts to occur when the current density reaches a certain value (critical current density, J_c), where the electric field becomes smaller than the critical electric field, and can be expressed as

$$J_c(\text{UTC-PD}) = qn_c v_e^s \qquad (4.10)$$

$$J_c(\text{pin-PD}) = qp_c v_h^s \qquad (4.11)$$

where v_e^s is the electron saturation velocity in InP and v_h^s the hole saturation velocity in InGaAs.

Because of the larger saturation velocity and smaller critical electric field for electrons, J_c can be much larger for the UTC-PD than that for a pin-PD. Figure 4.16 shows the calculated J_c as a function of V_b for a UTC-PD and a pin-PD. Here, we assumed the following values and device parameters: $N_D = 0$, $v_e^s = 4 \times 10^7$ cm/s, and $E_e^c = 10$ kV/cm for InP [45], $N_D = 0$, $v_h^s = 4.8 \times 10^6$ cm/s, and $E_h^c = 50$ kV/cm for InGaAs [41], and $W_C(\text{UTC-PD}) = W_A(\text{pin-PD}) = 300$ nm, $S = 52$ μm^2, and $R_L = 25\,\Omega$ for both types of device. Although the ratio of J_c for the UTC-PD and pin-PD depends on V_b and R_L, there is about a five-fold difference at $V_b = -1.5$ V, and the ratio further increases when the $W_C(=W_A)$ becomes larger. In reality, the charge distribution in the depletion layer is not uniform so that the output current saturation starts to occur at lower current density than in this calculation, especially with pin-PDs, resulting in a larger ratio of J_c. In addition, the pin-PD requires a certain negative bias voltage to work properly with the carrier velocity saturated, while the UTC-PD can be operated at zero or even positive bias voltages. A smaller W_C is further effective for expanding the operation area of the UTC-PD towards the positive bias voltage region. The experimentally obtained J_c deduced from the results shown in Figure 4.15 are also plotted in Figure 4.16 for comparison. The experimental results show fairly good agreement with the calculation. Although the model used in the above calculation includes several simplifications, it gives a reasonable measure of the critical current density for the UTC-PD and indicates again its superior high output current and low operation voltage features.

There is another factor that can influence the saturation of the output current. Under high-output current conditions, the operating point shifts considerably to the lower reverse bias voltage along the load line. This shift reduces the bandwidth of the PD through the increase of junction capacitance and the reduction of the electric field in the depletion region, resulting in a decrease of the maximum output current. The decreased electric field in the depletion layer also enhances the space charge effect, which also decreases the saturated output current. In addition, if the bias voltage is very small, the bias point can move into the forward biased condition along the load line and part of the output signal is clipped at the turn-on I–V curve of the PD [19]. Such an undesirable operation point shift can be avoided by properly increasing the reverse bias voltage when a very high output current is required.

For a high output current operation, some device parameters must be carefully optimized. Making the absorption area larger [57] and the collection layer thinner can increase the maximum output current [57], but doing so causes the CR-time limited bandwidth to deteriorate.

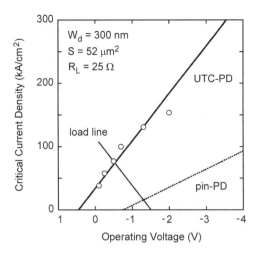

Figure 4.16 Calculated critical current density against operating voltage for a UTC-PD and pin-PD. Experimental results deduced from Figure 4.15 are also shown

Higher collection layer doping (either uniform [58] or non uniform [59]) is also advantageous. However, it may result in smaller junction breakdown voltage and/or increased junction capacitance. A more sophisticated design for the collection layer, which originates from the optimized collector layer structure of an HBT (or ballistic collection transistor (BCT) [28,29]), is also promising, and a high output current has also been demonstrated in a UTC-PD with a collection layer structure similar to the BCTs [60].

A device configuration for better heat sinking is also important, because the maximum operational current may be limited by thermal destruction [61]. In this regard, UTC-PD structure is advantageous, since the InP collection layer, on which the high electric field is applied, has a much larger thermal conductivity [62] than that of an InGaAs absorption layer [62], on which the high electric field is applied in the pin-PD structure. The difference in power dissipation tolerance between these two devices has also been experimentally demonstrated [63]. Other conventional techniques for heat sinking, such as the use of thick and wide-top metal electrodes, substrate thinning, and buried structure with epitaxial InP layers [64], should also be effective.

Figure 4.17 shows an example of a very-high output pulse photoresponse of a UTC-PD in a high-excitation condition [50]. Here, the same device as in Figure 4.15 was employed, with a bias voltage of -6 V and optical input energy of 8.7 pJ/pulse. The peak photocurrent reaches as high as 184 mA (corresponding to a current density of 350 kA/cm^2), while the FWHM stays at a relatively small value of 4.8 ps. The Fourier transform of this pulse response resulted in an f_{3dB} of 65 GHz. This peak photocurrent is about an order of magnitude larger than that obtainable with conventional pin-PDs with similar bandwidths. Figure 4.18 summarizes the results of the maximum peak output current against f_{3dB} for UTC-PDs, obtained by pulse photoresponse measurements, though the device and operational parameters are different for each device. As shown in this figure, the maximum output current and the bandwidth are in a trade-off relationship through device and operational parameters, such as absorption area and bias voltage. Nevertheless, the UTC-PD is considered to simultaneously provide a very high peak output current of about 100 mA with a very high f_{3dB} of 100 GHz.

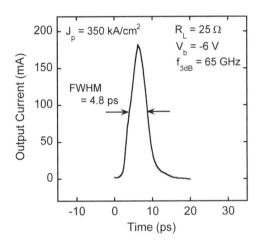

Figure 4.17 Pulse photoresponse of a UTC-PD at a very high output current conditions [50]. (Reproduced by permission of © 2001 IEICE)

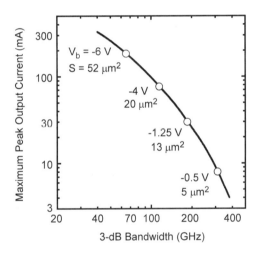

Figure 4.18 Maximum peak output current against 3-dB bandwidth for UTC-PDs obtained by pulse photoresponse measurements

4.2.3.7 Efficiency

Another important figure of merit is efficiency, which is directly related to absorption layer thickness. Although a thicker absorption layer is preferable for higher efficiency, the f_{3dB} basically decreases with increasing W_A. Thus, like the other structural parameters, the absorption layer thickness should also be optimized. To date, several types of PD structure have been developed, and some of them are beneficial for increasing the efficiency. Figure 4.19

schematically shows the classification of the PD structure in terms of the light illumination direction. The top-illuminated structure (No. 1) is a basic configuration and most widely used because it is advantageous for testing and assembling for low-cost modules, though it is not suitable for high-speed and high-current operation. The back-illuminated configuration (No. 5) has certain advantages over the top-illuminated one since it can use a light signal reflected at the surface electrode to increase the efficiency. The parasitic resistance can also be significantly reduced by making the contact area of the top electrode larger and by eliminating lateral current flow in the top contact layer [65]. Thus, this structure is also suitable for high-speed and/or high-current operation. Figure 4.20 shows the relationship between the 3-dB bandwidth and efficiency in back-illuminated UTC-PDs (open circles) [50]. A trade-off between f_{3dB} and efficiency through the absorption layer thickness is clearly seen.

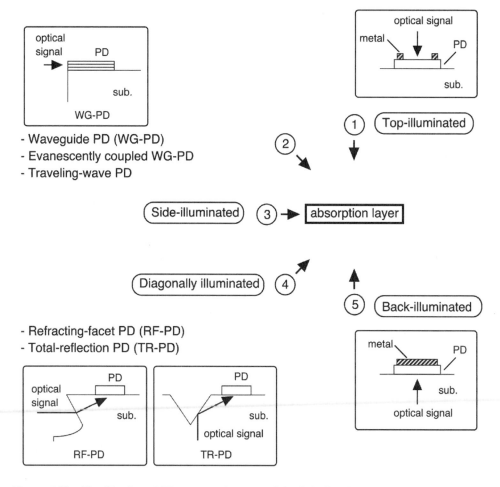

Figure 4.19 Classification of PD structure in terms of the light illumination direction. Examples of device structures with schematic cross-sectional views are also shown

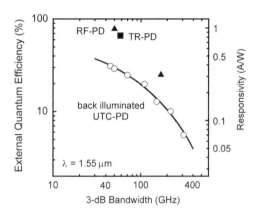

Figure 4.20 Relationship between 3-dB bandwidth and efficiency for several types of UTC-PDs [50]. (Reproduced by permission of © 2001 IEICE)

To increase the efficiency without sacrificing the high-speed characteristics, several types of structure have been proposed. One is a side-illuminated structure (No. 3), which increases the efficiency by increasing the light propagation length in the waveguide-structured absorption layer while maintaining the thin carrier traveling distance for the high-speed operation by making the directions of carrier transport and light propagation cross at a right angle [66]. One of the drawbacks of this structure is its relatively low maximum output current, which is due to the concentration of current at the irradiation edge and may cause a catastrophic degradation. The evanescently coupled structure [67] solves this problem by suppressing the current concentration at the illumination edge. Another promising solution for the waveguide structure is the traveling-wave electrode [68] or distributed velocity-matched structure [69], which increases the bandwidth by relaxing the influence of CR charging time, and to increase the output current by a proper distribution of the current flow. Some of these ideas, such as a waveguide structure [70], a velocity-matched distributed structure [71], and an evanescently coupled structure [72] have been applied to UTC-PD structure.

Another type of structure for enhanced efficiency is the diagonally illuminated structure (No. 2 or 4). Because there are practical problems in realizing No. 2, such as the total reflection of the input light on the device surface and the need for a specially designed optical module for the diagonal irradiation, No. 4 structure is realistic and frequently used. Good examples are the refracting-facet (RF) [73, 74] and total-reflection (TR) [75] structures, which utilize angled irradiation of the signal light onto the absorption layer to increase the light propagation length within it. In order to guide the input light diagonally, the RF-PD uses the refraction of the edge-illuminated light at an angled edge fabricated by using the material characteristic of InP substrate, in which the chemical etching spontaneously stops on the (111)A facet. Similarly, the TR-PD employs the total-reflection of back-illuminated light at a V-grooved (111)A facet mirror formed adjacent to the PD, which also uses the chemical etching nature of the InP substrate. Both structures are shown in the inset of Figure 4.19. In both cases, the PD itself has a simple double-mesa structure similar to that of the back-illuminated one. One thing we have to be careful about here is that, in principle, the efficiency depends on the polarization of the input light in the diagonal irradiation configuration, though the dependence can be drastically suppressed by optimizing the layer structure [74]. The results for an RF-UTC-PD (solid triangles) and a TR-UTC-PD (solid square) are also shown in Figure 4.20. The RF-PD achieved a responsivity of 1.0 A/W with an f_{3dB} of 50 GHz [44] and 0.32 A/W with an f_{3dB} of 170 GHz, while the values are 0.83 A/W with an f_{3dB} of 58 GHz for the TR-PD. These efficiencies are more than twice as large as those obtained by back-illuminated devices with similar f_{3dB} values.

A different approach to increasing the efficiency without sacrificing the device operation speed is to use a hybrid structure of a UTC-PD and a pin-PD, i.e., neutral and depleted absorption layers [76,77]. By optimizing the layer parameters, this configuration enables us to increase the total absorption layer thickness, while the carrier transport limited bandwidth stays almost the same [78]. However, this approach sacrifices the maximum output current to some extent because it partly uses hole transport in the device operation. Thus, there is a suitable area to which this hybrid structure should be applied. Another way is to hybrid the collection layer structure in combination with narrow-gap and wide-gap materials. In this way, the efficiency can be increased while maintaining the bandwidth by effectively utilizing a shorter hole traveling distance and a low junction capacitance [79]. This concept was originally proposed to improve the performance of a pin-PD [80]. A combination of this hybrid absorption/collection layer with the neutral absorption layer, with proper design considerations of the trade-offs among bandwidth, output current, and efficiency, would be a promising solution for practical applications.

4.2.3.8 Reliability

Reliability is also a very important issue for practical applications. To ensure the good reliability of a photodiode, a widely used standard stress test (Telcordia Generic Reliability Assurance Requirements, GR-468-CORE) has also been adopted to the UTC-PD. As an example, a bias-temperature stress test of UTC-PDs designed for 40-Gbit/s optical communications systems was performed at up to 240 °C with a bias voltage of −3 V [81]. The PD has a responsivity of 0.8 A/W and an f_{3dB} of 47 GHz at a photocurrent of 10 mA. The change in the dark current was very small over 1600 hours, and no failure was observed. Taking the results at 170, 200, and 240 °C with device numbers of 25, 25, and 24 for each condition into account, the failure rate was estimated to be smaller than 42 FIT at 25 °C. Here, the random failure mode with an activation energy of 0.35 eV, which is recommended by the International Electrotechnical Commission, and a confidence level of 60 % were assumed. This failure rate satisfies the general requirement for practical systems applications (less than 100 FIT). Because the UTC-PD is typically operated at a high current level, the long-term stability under an optical input stress is also of great interest. Thus, an optical input and bias stress test at a photocurrent of 10 mA with a bias voltage of −2 V at room temperature was also performed [82]. Here, the device had a responsivity of 0.35 A/W and f_{3dB} of about 100 GHz. Figure 4.21 shows the variation of the dark current against aging time during the stress test [2]. The dark current is less than 100 nA, and stays almost constant (except for the initial stage) for more than 8000 hours. This result indicates that UTC-PD is also reliable for relatively high optical input stress.

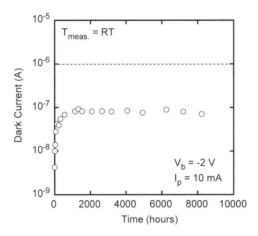

Figure 4.21   Variation of dark current against aging time during the optical input stress test [2]. (Reproduced by permission of © 2004 IEEE)

4.2.4 Photo Receivers

4.2.4.1 New Photo Receiver Configuration

A typical digital application of the UTC-PD is a photo receiver for ultrahigh bit-rate optical communications systems. Figures 4.22(a) and (b) show configurations of a new photo receiver with an optical amplifier and a UTC-PD, and a conventional one with a pin-PD and electrical post-amplifiers, respectively. Although a conventional photo receiver in long-haul systems also typically uses an optical amplifier in front of the photodiode, an electrical post-amplifier is needed in order to make the output electrical signal large enough to drive the decision circuit properly because of the low output current from a conventional high-speed pin-PD. In contrast, the high output capability of the UTC-PD used in combination with an optical amplifier (Figure 4.22(a)) makes it possible to drive the decision circuit directly without post-electrical-amplifiers [13]. Such a configuration can eliminate drawbacks in using the electrical amplifiers (limited bandwidth, additional noise, increased power consumption, etc.), and provides several advantages, such as wider bandwidth, a simpler system, and better sensitivity for systems operating at 40 Gbit/s or higher [14, 16].

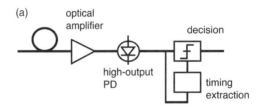

Figure 4.22   (a) A new photo receiver configuration consisting of an optical pre-amplifier and a high-output PD, and (b) a conventional one consisting of a pin-PD and electrical post-amplifiers

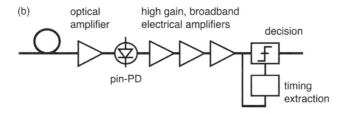

Figure 4.22   (*continued*)

4.2.4.2 Photodiode Module for High-bit-rate Systems

For the photo receiver application, a UTC-PD module with a 3-dB bandwidth of over 65 GHz is now commercially available. Although the UTC-PD chip can be operated at even higher frequencies, the electrical connection in the current hybrid assembly scheme (V-connector technology) limits the operation at frequencies higher than 75 GHz. One way to overcome this limitation is to use a 1-mm connector [83], which has a bandwidth of 110 GHz. Figure 4.23 shows a photograph of a UTC-PD module with a 1-mm connector [84]. A UTC-PD with an f_{3dB} of 93 GHz and a responsivity of 0.21 A/W was monolithically integrated with a biasing circuit consisting of a capacitor and a matched resistor (50 Ω) and electrically connected to the 1-mm connector by means of a micro-strip line fabricated on quartz substrate (for low dielectric loss). Then, the PD was optically coupled with an optical fiber by using a two-lens system. The optical parts were fixed onto the package using a YAG laser welder.

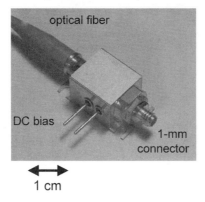

Figure 4.23   Photograph of a UTC-PD module with a 1-mm coaxial connector output [84]. (Reproduced by permission of © 2004 IET)

Figures 4.24(a) and (b) show the output waveforms of the fabricated UTC-PD module for 100- and 160-Gbit/s return-to-zero (RZ) signal [84], observed with an 80-GHz-bandwidth sampling oscilloscope (Agilent 86116B) at a bias voltage of −2 V. Clear eye openings with very high eye amplitudes of 0.5 V at 100 Gbit/s and 0.3 V at 160 Gbit/s were obtained. To our knowledge, this is the first demonstration of a photo receiver operation for a 160-Gbit/s RZ signal by a PD module operating at 1.55 μm. For the 100-Gbit/s case, timing jitter was less than 400 fs and no distinguishable phase delay was observed. On the other hand, for the 160-Gbit/s case, degradations in the eye amplitude and in the waveform (to be NRZ-like), including a signal phase delay, are observed. These degradations are attributed to several factors, such as the insufficient bandwidths of the fabricated module, the sampling head (80 GHz), and 1-mm connector (110 GHz). Thus, for higher bit-rate applications, further progress in device assembly technology, connector technology, and evaluation equipment technology are necessary. Another way to overcome these limitations is to integrate the UTC-PD with an electronic circuit so that the demultiplexed signals are output from the OEIC without suffering from the limitations of the electrical signal connection.

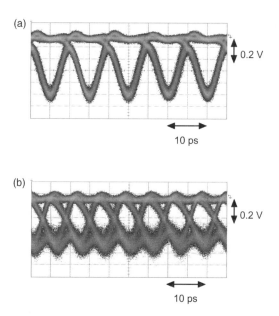

Figure 4.24   Eye diagrams for 100-(a) and 160-Gbit/s (b) RZ optical signal inputs received using the 1-mm connector UTC-PD module [84]. (Reproduced by permission of © 2004 IET)

Figure 4.25 shows the eye amplitude against the photocurrent for several bias voltages at 100 Gbit/s [84]. The eye amplitude increases linearly with increasing photocurrent until it saturates. For the bias voltage of −3 V, the linearity is maintained for eye amplitudes of over 0.5 V, and the maximum value exceeds 0.8 V, which is sufficient to directly drive the digital circuit at 100 Gbit/s without employing electronic amplifiers. A similar tendency was also seen at 160 Gbit/s with a linear output voltage of more than 0.5 V, and the maximum value exceeded 0.6 V. This module was also operated at zero bias voltage with an output voltage of over 200 mV at up to 160 Gbit/s.

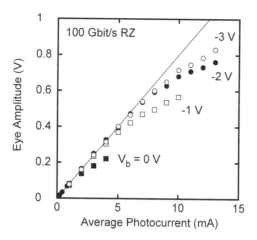

Figure 4.25 Eye amplitude against average photocurrent at 100-Gbit/s RZ signal input for several bias voltages [84]. (Reproduced by permission of © 2004 IET)

4.3 Concept of a New Opto-electronic Integrated Device

4.3.1 Importance of High-output PDs

As explained in the previous section, signal processing by the conventional OE-based device requires amplification of electrical signals because the output voltage of a high-speed pin-PD is usually not high enough to directly drive optical and electrical devices, such as LDs, optical modulators and digital circuits. Especially for very high speed operation, a trade-off between operation speed and output current in the conventional pin-PD makes the available output current much smaller. Thus, broadband electrical amplifiers are frequently used in optical transmitters, optical modulators, and photo receivers. However, the maximum operation frequency of the baseband electrical amplifiers reported so far is around 100 GHz [85, 86], and progress in their performance is still needed for practical use in ultrafast optical signal-processing systems operating at 160 Gbit/s or higher. One way to overcome such a limitation is to achieve the function of electrical amplifiers by replacing them with broadband optical amplifiers (typically an optical fiber amplifier) and a high-speed, high-output PD, which directly drive the following optical or electrical device. Such a configuration provides several advantages, such as wider bandwidth, better sensitivity, and simpler electrical signal connection. For this purpose, the UTC-PD is one of the best choices for O/E conversion device (or opto-electronic driver), since it offers high speed and high-output current simultaneously, as described in the previous section. Monolithic integration is also an effective way to make the best use of each elemental device. There are several promising integrated devices that effectively use the features of the UTC-PD, not only as an all-optical signal processing device but also as receiver OEICs for the operation at speeds of over 100 Gbit/s.

4.3.2 Monolithic Digital OEIC

The integration of optical and electrical devices is a traditional way to create novel functions and improve device performance. The use of a UTC-PD instead of a pin-PD can eliminate speed-limiting electrical amplifiers and thus extend circuit bandwidth. There are several promising combinations for such direct-drive type integration. The monolithic integration of a UTC-PD with an electrical logic circuit can provide high-speed digital OEICs for operation at 40 Gbit/s or above [3, 87, 88]. Error-free photo receiver operation at 40 Gbit/s with a sensitivity of −27.5 dBm has been achieved with an opto-electronic decision circuit that integrates a UTC-PD and a high-electron-mobility transistor (HEMT) circuit (with 172 active electron devices) [87]. Optical-clock-frequency-divider OEICs for operations of up to 100 GHz have also been demonstrated by integrating a UTC-PD and HEMT-based [89] or HBT-based [6] circuit.

The monolithic integration of a UTC-PD with ultrafast resonant tunneling diodes (RTDs), which have a switching time of a few picoseconds with low switching power, has been applied to an opto-electronic demultiplexer (80 to 40 Gbit/s) [90] and a 80-Gbit/s delayed flip-flop circuit [3]. A decision circuit operating at 40 Gbit/s [91] and an opto-electronic clock recovery circuit operating at 46.2 Gbit/s [88] with the same configuration have also been reported. These novel circuits are promising for high-speed, low-power, opto-electronic logic gates.

4.3.3 Monolithic PD-EAM Optical Gate

A novel combination of unique elemental devices may lead to an effective fusion of the superior performance of each device, resulting in a new class of integrated device. A good example is the combination of a UTC-PD with an optical modulator, such as an EAM, corresponding to configuration (iii) in Figure 4.1(b). This combination provides an O/E/O-type optical gate without speed-limiting electrical amplifiers. Such an idea was first demonstrated using a simple 'discrete module combination', and a 40 to 10-Gbit/s DEMUX [92] and 10-Gbit/s 2R signal reshaper [10] have been reported. In that configuration, however, the electrical interconnections between each elemental device, including the connectors and electrical waveguides used in the device packaging, limited the maximum operation speed. In addition, a CR time constant of the lumped-element EAM also limited the operation speed.

To overcome the speed problem of the discrete configuration mentioned above, NTT Laboratories has developed (in 1998) a PD-EAM optical gate that monolithically integrates a UTC-PD and a traveling-wave-electrode EAM (TW-EAM) [4]. The main feature of the monolithic PD-EAM optical gate is that the EAM is directly driven by a high voltage signal generated by the UTC-PD without an electrical amplifier. This greatly improves the bandwidth of the optical gate. In addition, by effectively exploiting the nonlinear extinction characteristics of the EAM, the gate opening time can be considerably reduced. Furthermore, the use of a TW electrode structure in the EAM is effective for eliminating the influence of the CR time constant [93,94]. Finally, monolithic integration of each element eliminates the degradation of propagating electrical waveforms, which is a serious problem in the discrete-module-connection configuration. By combining all of these features, a broadband, simple, and compact optical gate can be realized. This is a good example of how the excellent high-speed and high-output characteristics of the UTC-PD can open the way to a new class of opto-electronic integrated devices, which can be operated at speeds beyond those possible with conventional OEIC technology and provide the various fundamental functions required in optical-signal-processing systems.

4.4 PD-EAM Optical Gate Integrating UTC-PD and TW-EAM

4.4.1 Basic Structure

Figure 4.26(a) shows drawing of the monolithic PD-EAM optical gate, and Figure 4.26(b) shows its circuit diagram. The device consists of a back-illuminated UTC-PD, a TW-EAM, two DC-bias capacitors, a terminal resistor (R_T), and waveguides (either CPWs or microstrip lines (MSLs)) connecting with key elements (UTC-PD, TW-EAM and terminal resistor). All of these elements can be fabricated by a sequential process after the epitaxial growth of the TW-EAM and UTC-PD layers. In the configuration shown in Figure 4.26(a), the anode of a UTC-PD is directly connected to the anode of the TW-EAM by a waveguide. Such a config-uration is possible because the UTC-PD can supply sufficient voltage to drive the TW-EAM. The other side of the TW-EAM anode is connected to the terminal resistor by another wave-guide to avoid signal reflection. The bias voltages are applied to the UTC-PD and TW-EAM independently by two biasing circuits. Here, a positive applied voltage to each bias terminal provides a negative bias voltage to each device. This configuration makes the bandwidth of the PD-EAM much higher than that possible using external bias tees. Each element is arranged to be close enough (but not too close) to avoid degradation of the propagating signal wave-form, standing-wave generation, and interference of active devices (UTC-PD and TW-EAM). In this regard, monolithic integration is essential. The optical input signal from an optical fiber is coupled to the TW-EAM waveguide from one end by a focusing lens, and the output

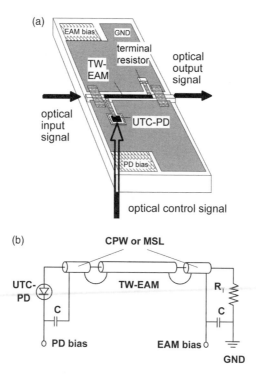

Figure 4.26 (a) Schematic view and (b) circuit diagram of the monolithic PD-EAM optical gate

optical signal from the other end of the TW-EAM waveguide is coupled to another optical fiber. Another optical input signal, which controls the operation of the TW-EAM, is coupled to the UTC-PD from the back of the chip. For gating operations, the TW-EAM is biased so that the waveguide is not transparent for 1.55-μm light signals. With optical input (control signal) to the UTC-PD, the generated photocurrent provides a positive bias voltage to the TW-EAM, making the waveguide transparent for the input optical signal. Thus, the PD-EAM functions as a transmission-type (normally closed) optical gate and outputs signals only when the synchronized control signal is input to the UTC-PD. Similarly, an inverter-type (normally open) optical gate can be constructed by simply changing the output polarity of the UTC-PD by connecting the cathode of the UTC-PD to the anode of the TW-EAM. For this operation, the TW-EAM should be biased so that the waveguide is transparent to 1.55-μm light signals and not transparent when the control signal is on. These two types of optical gate can be fabricated simultaneously on the same wafer without changing or adding any process steps. Thus, by properly combining these PD-EAMs with optical waveguides, novel functional optical integrated devices with higher complexity can also be constructed. For this purpose, a waveguide-type UTC-PD [70, 71] would be beneficial for putting the optical paths on the same plain.

Because there is no high-speed electrical signal interface outside the chip, the PD-EAM is free from the 50-Ω impedance-matching restriction. This is an important feature of the PD-EAM, allowing us to use smaller characteristic impedance for the TW-EAM waveguides, and terminal resistor to relax the influence of the CR time constant for the UTC-PD. On the other hand, the use of smaller characteristic impedance sacrifices the output signal voltage from the UTC-PD and requires a higher output current from (higher optical input power to) the UTC-PD. Thus, design optimization will be necessary in order to achieve the best PD-EAM performance under practical operation conditions.

4.4.2 Design

The operation speed of a PD-EAM is determined by the photoresponse of the UTC-PD and the electro-optic response of the TW-EAM. Smooth propagation of the electrical signal from the UTC-PD to the terminal resistor through the EAM and interconnections is also important for ultrafast response of the PD-EAM. In this subsection, guidelines for designing the key elements of the PD-EAM are described.

4.4.2.1 UTC-PD

In order to drive the EAM directly, high output voltage from the UTC-PD is basically required. However, as the output current density of a UTC-PD increases, its bandwidth tends to show a decrease associated with the output saturation. Thus, a relatively large junction area is necessary for a high output current, as described in the previous section. On the other hand, it lowers the bandwidth associated with the CR time constant. Therefore, UTC-PD parameters have to be properly determined to obtain wide bandwidth and high output current simultaneously so that the total performance of the PD-EAM is maximized. To accomplish this, the characteristics of the UTC-PD, the TW-EAM, and the electrical waveguide must be taken into account. A benefit of the smaller characteristic impedance allowed for the PD-EAM structure is that a relatively large-area UTC-PD for a higher output current can be used without seriously sacrificing its CR-time-limited bandwidth. For example, an absorption layer thickness of 100 nm, a collection layer thickness of 300 nm, an absorption area of $70\,\mu m^2$, and a load resistance of $15\,\Omega$ will result in an f_{3dB} of over 150 GHz, with a peak output current of more than 140 mA, which can generate a driving voltage of more than 2 V for the EAM.

4.4.2.2 TW-EAM

The bandwidth of a TW-EAM is basically limited by the propagation-velocity mismatch between the optical and electrical signals. Optical responses $R(\omega)$ of the TW-EAM are expressed as:

$$R(\omega) = \frac{1 - \exp[-j\omega(v_0 - v_e)v_e^{-1}\tau_o]}{-j\omega(v_0 - v_e)v_e^{-1}\tau_o} \tag{4.12}$$

where v_o and v_e are the propagation velocities of optical and electrical signals, respectively. Here, τ_o is the propagation time of optical signal in the EAM and is expressed as:

$$\tau_o = L_{EAM}/v_o \tag{4.13}$$

where L_{EAM} is the active waveguide length of the EAM. From Equation (4.12),

$$f_{3dB} = \frac{1}{\sqrt{2}\pi(v_o - v_e)v_e^{-1}\tau_o} \tag{4.14}$$

The bandwidth inversely depends on L_{EAM}, and thus a smaller L_{EAM} is preferable for higher f_{3dB}. On the other hand, we have to maintain the required extinction ratio of the EAM by keeping a sufficient waveguide length. So, a realistic design of L_{EAM} is a compromise between the bandwidth and EAM extinction ratio with those required for a specific application.

For example, f_{3dB} for a L_{EAM} of $200\,\mu m$, which gives an extinction ratio of about 30 dB, is calculated to be 110 GHz, assuming $v_o = 1 \times 10^{10}$ cm/s and $v_e = 0.5 \times 10^{10}$ cm/s as values in a realistic TW-EAM structure.

4.4.2.3 Impedance Matching

The operation speed of the PD-EAM is also limited by reflection of the electrical signal propagating within it. In the electrical paths of the PD-EAM, there are several structural discontinuities where the signal reflection can occur due to impedance mismatching. For the smooth propagation of electrical signals, which results in a smooth response waveform, impedance matching between each element including the signal lines (waveguides) is very important.

Figure 4.27 shows the characteristic impedance of a fabricated TW-EAM obtained from the electrical small-signal response measurement. The TW-EAM has a low impedance of about $15\,\Omega$. Since there is no electrical interface outside the chip, no 50-Ω electrical design is required in a monolithic PD-EAM optical gate. This means that we can use this low-characteristic impedance for all of the elements, including the waveguide for the PD-EAM. However, on the semiconductor substrate, it is generally difficult to obtain such low characteristic impedance with a CPW, which is widely used in high-speed electrical circuits. A solution is to use a thin-film MSL. A low characteristic impedance can be obtained by making the strip a lot wider than the thickness of the insulator between the strip and the ground [95].

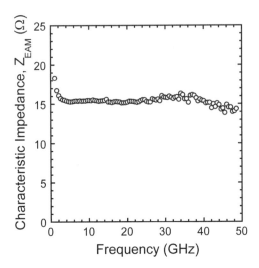

Figure 4.27   Characteristic impedance of a fabricated TW-EAM obtained from electrical small-signal response measurement

Figure 4.28 shows the calculated and measured characteristic impedances of a thin-film MSL with a benzocyclobutene (BCB) insulator layer ($\varepsilon_r = 2.65$) between the strip and the ground plate. A characteristic impedance of 15 Ω, which matches that of the TW-EAM, can be obtained with a realistic strip width of 32 μm and a BCB layer thickness of 2.5 μm. Since the terminal resistor with a resistance of 15 Ω can be conventionally fabricated, impedance matching between all the elements in the PD-EAM can be accomplished by adopting this low characteristic impedance.

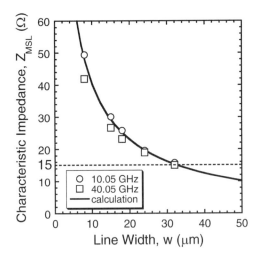

Figure 4.28   Calculated and measured characteristic impedance of a thin-film MSL with a benzocyclobutene (BCB) insulator layer between the strip and ground plate

4.4.3 Optical Gating Characteristics of PD-EAM

4.4.3.1 Reduction of Gate Opening Time by Nonlinear Extinction Characteristics of the EAM

The operation speed of a PD-EAM is mainly determined by the photoresponse of the UTC-PD and the electro-optic response of the EAM. Thus, employing a UTC-PD and EAM with faster responses is a basic way to obtain shorter gate opening times in the PD-EAM. The gate opening time can also be shortened by effectively using the nonlinear extinction characteristics of the EAM [96]. Figure 4.29 schematically shows how to reduce the gate opening time. When a large reverse bias voltage for off-state (nontransparent state) is applied to the EAM and the UTC-PD outputs a sufficient voltage to turn the EAM on, only a part of the voltage near the summit of the electrical output pulse is used for gate opening. The optical output pulse of the PD-EAM then becomes narrower than that of the electrical output pulse of the UTC-PD. This operation mode is useful for applications that require a short gate-opening time with a relatively low repetition rate, such as DEMUX, optical sampling, and short optical pulse generation. A drawback of this operation mode is that it requires a higher optical input power, and thus a higher output current from the UTC-PD, which may limit the degree of the pulse width reduction.

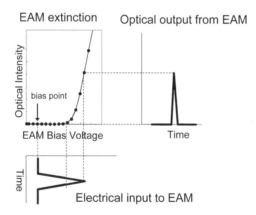

Figure 4.29 Schematic diagram showing how the gate opening time is reduced by nonlinear extinction characteristics of the EAM

4.4.3.2 Rectangular-like Gating Waveform

The gating waveform of the PD-EAM can be modified by properly choosing its operation conditions, and the shape of the waveform greatly affects the PD-EAM performance. In particular, the generation of a rectangular-like gating waveform is important for improving its characteristics. Figure 4.30 schematically explains how to obtain a rectangular-like gating waveform. When the optical input is very strong, the electrical output from the UTC-PD saturates. Even in this case, rise and fall times of the electrical output pulse are still very small [15] as shown in the upper part of the diagram. Using this unique output saturation behavior of the UTC-PD together with the nonlinear extinction (saturated transparency) characteristic of the EAM (lower left), the PD-EAM can produce a rectangular-like gating waveform (lower right).

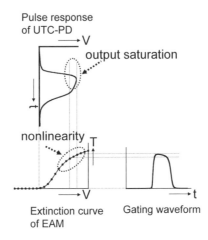

Figure 4.30 Schematic diagram showing how the rectangular-like gating waveform is established in the PD-EAM

Such a rectangular-like gating waveform is beneficial for various applications since it can relax the timing-jitter penalty in the input optical signals. Figure 4.31 shows a retiming operation using a PD-EAM as an example. Even when the input data pulses for the UTC-PD experience a timing shift against the clock pulses for the EAM, the PD-EAM can properly gate the clock pulses within the timing margin corresponding to the width of the rectangular-like gating waveform, as shown in the bottom diagram. This is another important feature of the PD-EAM for its application to practical systems.

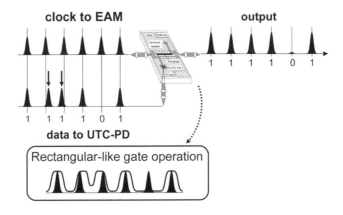

Figure 4.31 Schematic diagram showing how the rectangular-like gating waveform of the PD-EAM is effective for providing a timing margin in the gating operation

4.4.4 Fabrication

The PD-EAM fabrication process begins with the epitaxial growth of the stacked EAM and UTC-PD layers. Figures 4.32(a) and (b) show the layer structures. These layers are grown on a (100) oriented 2-inch S.I. InP substrate by MOCVD in two growth steps. The EAM layers are grown first, and then the UTC-PD layers are grown on top of them after proper surface treatment before the regrowth. In the EAM structure, we used six-period, strain-compensated, InAlGaAs/InAlAs multiple quantum well (MQW) layers [97], which were designed to minimize the polarization dependence of the EAM. For the UTC-PD structure, the doping and thickness of the p-InGaAs absorption and the n-InP collection layers are the design parameters, and are varied depending on the performance target.

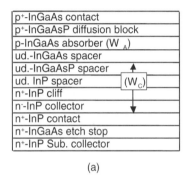

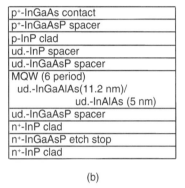

(a)   (b)

Figure 4.32   Epitaxial layer structures of a UTC-PD and an EAM used for the PD-EAM

Figures 4.33(a)–(d) show the fabrication sequence of the monolithic PD-EAM optical gate. The process starts with the fabrication of the UTC-PDs (Figure 4.33(a)). Conventional wet chemical etching and metal lift-off techniques are used for the fabrication process as described in the Section 4.2.2. Next, wet chemical etching and metal lift-off processes fabricate the EAM high-mesa waveguide structure after removal of the UTC-PD layers on top (Figure 4.33(b)). The typical width of the TW-EAM mesa is 3 μm. In this EAM fabrication process, the UTC-PD is isolated by deep mesa etching down to the SI substrate simultaneously. Then, the MIM bias capacitors and the terminal resistor are formed using silicon dioxide as an insulator and tungsten nitride silicide as a resistor, respectively (Figure 4.33(c)). The terminal resistor is fabricated on the SI substrate with a silicon dioxide insulating layer for better isolation. Then, all the elements are embedded using BCB for planarization. Finally, an interconnecting metal layer is deposited on the BCB layer (Figure 4.33(d)). The backside of the wafer is mirror polished and AR coated, and both sides of the EAM waveguide are also AR coated after the wafer has been cleaved.

Figure 4.34 is a micrograph of a fabricated PD-EAM chip [8]. The chip size of this device is 1 mm × 0.4 mm. The EAM waveguide has a 200-μm long active part and two 100-μm long passive parts. These passive waveguides (fabricated by shortening the p and n contact layers) are placed on either side of the active waveguide to make the chip large enough to arrange necessary elements on it, and for ease of chip handling. The input optical signal to the EAM waveguide comes from the upper side of the device in this picture, and the output signal comes out from the bottom side of the chip. The input optical signal to the UTC-PD comes from the back in this picture.

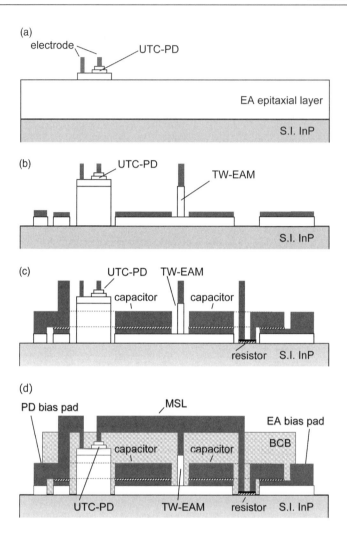

Figure 4.33 Fabrication sequence of the PD-EAM. (a) Fabrication of the UTC-PD; (b) fabrication of the TW-EAM; (c) fabrication of the MIM capacitor and terminal resistor, and (d) planarization and formation of interconnections between elements

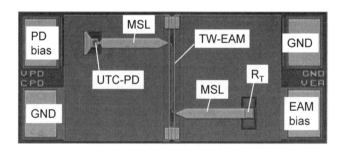

Figure 4.34 Micrograph of a fabricated PD-EAM [8]. (Reproduced by permission of © 2003 IET)

For practical use with ease of handling, the device has to be packaged. Thus, we also fabricated a three-optical-port module for the monolithic PD-EAM optical gate as shown in Figure 4.35 [98]. This is very compact, the size being the same as that of a conventional EAM module [99]. A Peltier device and thermistor were also assembled in the module to stabilize the device characteristics. This packaging also provides us with an important experimental environment with stable optical alignments, easy electrical connection, and easy device temperature control. The optical coupling efficiencies were almost identical to those of our specially designed probing stage, with three optical input ports, and were stable from 0 to 85°C. The dynamic characteristics of the device measured against a packaged device were also identical to those obtained by chip measurements on the probing stage. Hence, this module has been frequently used in measurements.

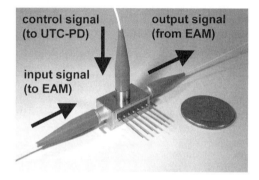

Figure 4.35 Photograph of a compact PD-EAM module with three optical ports [98]. (Reproduced by permission of © 2006 IEEE)

4.4.5 Gating Characteristics

To examine the fundamental performance of the PD-EAM, gating characteristics were evaluated in detail. In applying a PD-EAM to optical signal processing, important figures of merit are the gate opening time (pulse width), extinction ratio, and gating profile. To characterize them, we performed both static and pump-and-probe measurements. The layout of the pump-and-probe measurement system is illustrated in Figure 4.36. A 1.55-μm optical pulse generated by a mode-locked fiber ring laser (ML-FL) had an FWHM of 270 fs and a repetition rate of 32.6 MHz. The static extinction ratio of the EAM measured by applying DC bias voltage and a CW optical input of $+4$ dBm was as large as 35 dB for a voltage swing of 2 V. The total transmission loss was about 7 dB, including the fiber-to-EAM coupling loss.

Figure 4.37 shows the variation of the gating waveforms of the PD-EAM at EAM bias voltages (V_{EAM}) of -2.5, -2.75 and -3.0 V under a fixed PD optical input of 5 pJ/pulse. The FWHM decreases with increasing V_{EAM}, and a minimum FWHM of 2.1 ps was obtained with an extinction ratio of 14 dB. This reduction in FWHM is basically due to the nonlinear extinction characteristics of the EAM, as explained in Section 4.4.3.1. A good extinction ratio of 25 dB with an FWHM of 2.8 ps was also achieved at an applied voltage to the EAM of -2.5 V. The PD-EAM photoresponse also changes when the PD input level (P_{in}) is varied. Figure 4.38 summarizes variations of FWHM and the extinction ratio against V_{EAM} for several

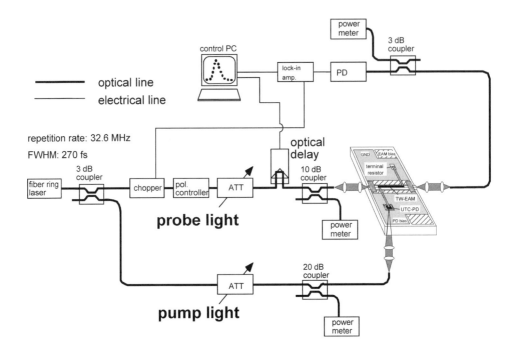

Figure 4.36   Experimental set-up for pump and probe measurement

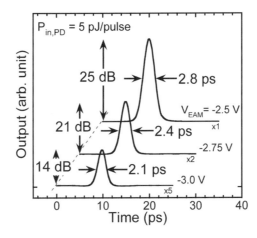

Figure 4.37   Variation of gate opening (pulse photoresponse) curves of a PD-EAM at EAM bias voltages of −2.5, −2.75, and −3 V under a fixed PD optical input of 5 pJ/pulse

P_{in} values. As P_{in} rises from 2 to 10 pJ/pulse, the FWHM increases and the extinction ratio rises significantly. For example, when V_{EAM} is fixed at −2.5 V, FWHM increases from 2.2 to 4.2 ps and the extinction ratio rises from 15 to 29 dB. The increase of the FWHM with increasing input power is basically attributed to the broadening of the electrical signal from the UTC-PD associated with the output saturation, while the increase of the extinction ratio with increasing

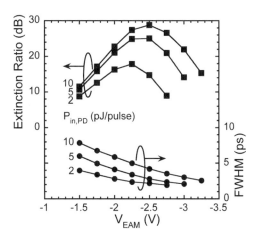

Figure 4.38   Dependence of gate opening time (FWHM) and extension ratio on the EAM bias voltage in a PD-EAM

optical input is due to the increased UTC-PD output voltage, which drives the EAM. At higher V_{EAM}, the extinction ratio decreases because the EAM bias voltage becomes much lower than its threshold voltage for the transmission, so that the applied voltage to the EAM becomes insufficient to fully open the EAM gate.

The minimum FWHM observed in this experiment was 1.9 ps. This value was determined primarily by the electrical pulse profile from the UTC-PD and the nonlinearity of the EAMs extinction characteristics. Another important factor responsible for limiting the minimum gate opening time is the velocity mismatch as described in Section 4.4.2.2. Along the active wave-guide ($L_{EAM} = 200 \,\mu m$), the traveling time difference between the optical and electrical signals is as large as 1 ps. However, the interaction between optical and electrical signals is rather localized on the light-input portion of the EAM. Therefore, total net traveling time difference may not directly affect the FWHM.

In actual signal-processing systems, inter-channel cross-talk is also an important issue. To characterize the gating characteristics of the PD-EAM for consecutive pulses with a very short period, we evaluated the optical output signal waveforms by using a cross-correlation technique. Figure 4.39 shows the experimental set-up [100]. Input data pulses (four pulses) fed into the EAM with a corresponding data rate of 500 Gbit/s (2 ps intervals) were prepared by multiplexing a 1.541-μm optical pulse generated by a passive ML-FL. The same initial pulses from the ML-FL, which had a FWHM of 600 fs, a wavelength of 1.541 μm and a repetition rate of 52 MHz, were provided to the UTC-PD and the cross-correlator as clock and reference pulses, respectively. The optical energy of the input signal fed to the EAM was fixed at 25 fJ/pulse. The EAM bias terminal voltage for the transmission and gate-off states were 0 and 2.2 V, respectively. The PD bias terminal voltage was kept at 3.5 V. Figure 4.40 shows (a) the input data pulses observed through the EAM at the transmission state and (b) each output pulse for each time slot (No. 1 – No. 4) after the DEMUX at a PD input (clock) pulse energy of 5 pJ/pulse [100]. Every input data pulse is clearly demultiplexed by adjusting the timing between the input data pulses and the clock pulse. This is the fastest DEMUX operation ever achieved for an OE-based device.

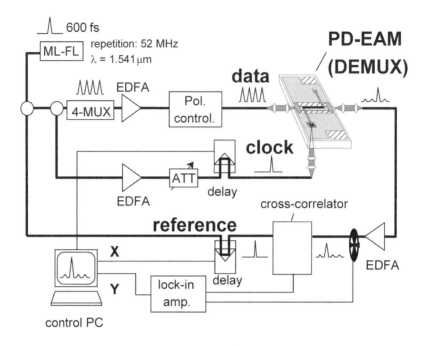

Figure 4.39 Experimental set-up for characterizing the gating operation of a pulse train by a PD-EAM at a speed corresponding to 500 Gbit/s [100]. (Reproduced by permission of © 2004 IET)

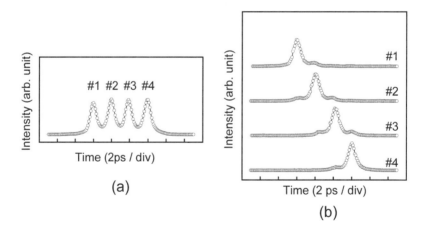

Figure 4.40 (a) Input data pulses with 2-ps intervals (corresponding to a data rate of 500 Gbit/s), and (b) output (demultiplexed) data pulses [100]. (Reproduced by permission of © 2004 IET)

In characterizing inter-channel cross-talk, we defined the on/off ratio as the peak intensity ratio between the on-level of pulse No. 2 and the off-levels of the other pulses (Nos 1, 3, and 4). Figure 4.41 shows the on/off ratio of the PD-EAM gating under 500-Gbit/s DEMUX operation [100]. A good on/off ratio against the adjacent pulses (No.2 / No.1 or No.2 / No.3) of 9 dB is obtained at a PD input pulse energy of 5 pJ/pulse. The on/off ratio decreases for the adjacent pulses at larger input energies than 5 pJ/pulse. This is mainly caused by the pulse width increase of the UTC-PD electrical output due to the output saturation. However, for pulses other than the adjacent ones, the on/off ratio (No.2 / No.4) is as high as 19 dB at a PD input pulse energy of 5 pJ/pulse. Thus, the on/off ratio could be further improved by reducing the pulse width of the UTC-PD output.

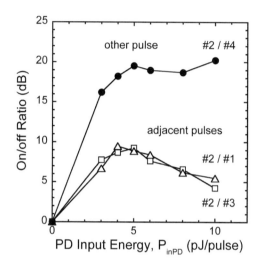

Figure 4.41   Measured on/off ratios against PD input (clock) pulse energy under 500-Gbit/s DEMUX with PD-EAM optical gate [100]. (Reproduced by permission of © 2004 IET)

These results demonstrate that the monolithic PD-EAM optical gate has a high extinction ratio, a very small FWHM, and low inter-channel cross-talk, and is suitable for application to ultrafast all-optical signal processing.

4.4.6 Applications for Ultrafast All-Optical Signal Processing

Figure 4.42 shows a diagram of a simplified optical transmission system and the required functions at each point. To construct such an ultrafast all-optical signal processing system, key devices with various functions have to be developed in a practical form. For this requirement, the monolithic PD-EAM optical gate can be used to achieve several functions required in such a system. In this section, several promising applications of the PD-EAM are described.

4.4.6.1 DEMUX

DEMUX is an indispensable function in ultrafast optical-signal-processing systems. Although a combination of a high-power electrical amplifier and an EAM have demonstrated DEMUX at a base data rate of up to 160 Gbit/s [101, 102], DEMUX in the optical signal stage is considered to be required when the data rate is beyond 160 Gbit/s. The PD-EAM functions as optical DEMUX

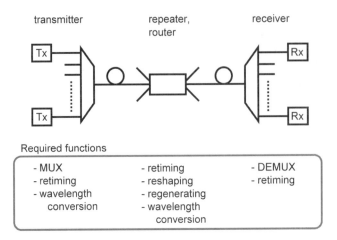

Figure 4.42 Schematic configuration of a simplified optical transmission system and the required functions

when an optical clock signal with a $1/n$ frequency of the data rate is synchronously supplied to the UTC-PD. Figure 4.43 shows the experimental set-up for characterizing DEMUX operation with a monolithic PD-EAM optical gate at data rates up to 320 Gbit/s [8]. The DEMUX consists of a PD-EAM, two Er-doped fiber amplifiers (EDFAs), and a variable optical delay line. The data and clock signals were prepared using a 1.55-μm optical pulse stream generated by an active ML-FL with a repetition rate of 9.95328 GHz and a pulse width of 800 fs. The data signals of up to 320 Gbit/s were prepared by multiplexing an optical 10-Gbit/s pseudo-random bit

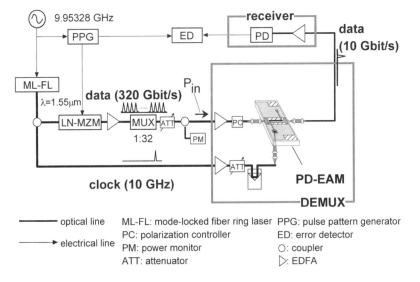

Figure 4.43 Experimental set-up for characterizing 320-Gbit/s DEMUX with the PD-EAM optical gate [8]. (Reproduced by permission of © 2003 IET)

sequence (PRBS) with a pattern length of 2^7-1. The same initial pulse stream from the ML-FL was also provided to the PD-EAM as a 10-GHz clock signal. Received optical power, P_{in}, is defined as the input to the EDFA in front of the EAM. The receiver sensitivity of this set-up without MUX and DEMUX (10-Gbit/s back-to-back) was -36.5 dBm. Applied voltages to the PD and EAM bias terminals, V_{PD} and V_{EAM}, were 4.0 and 2.0 V for 160-Gbit/s DEMUX and 4.0 and 2.4 V for 320-Gbit/s DEMUX, respectively.

Figures 4.44(a) and (b) show cross-correlation traces of input (320-Gbit/s) and output (demultiplexed 10-Gbit/s) data signals, respectively [8]. The pulse energies of input data and clock signals were 25 fJ/pulse and 10 pJ/pulse, and the corresponding average powers were 6 and 20 dBm, respectively. The pulse width after multiplexing was 1 ps. The traces indicate that the 320-Gbit/s input data is clearly demultiplexed by properly adjusting the timing between the input data signal and the clock signal. The on/off ratios, which are defined as the peak intensity ratio between the on- and off-levels, are 14 dB for the adjacent channels and 28 dB for the furthest channel between two on-channels. For the other channels, the on/off ratio behaved in the same way as it did for the furthest channel. Figure 4.45 shows the measured bit error rate (BER) for 320-Gbit/s to 10-Gbit/s DEMUX together with that for the 160-Gbit/s data input [8]. Error-free operations for 320-Gbit/s DEMUX was achieved with a receiver sensitivity of -18 dBm at a BER of 10^{-9}. This is the highest bit rate ever achieved for error-free operation using an OE-type optical gate. These results clearly demonstrate that the PD-EAM optical gate has a high potential for use in ultrafast optical signal-processing systems.

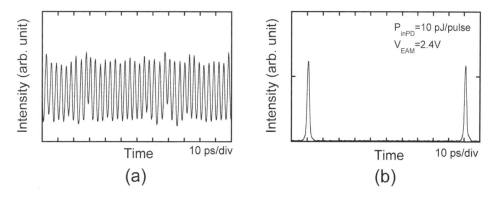

Figure 4.44 Waveforms for (a) 320-Gbit/s input data signal, and (b) 10-Gbit/s output (demultiplexed) data signal [8]. (Reproduced by permission of © 2003 IET)

For the actual demultiplexing function in the system with the PD-EAM, an optical clock recovery device is also required. For this purpose, we have developed an optical clock recovery device that uses a regeneratively mode-locked laser diode (MLLD-OCR) [103–105] and combined it with the PD-EAM to construct a receiver for optical time division multiplexing (OTDM) signals [106]. Figure 4.46 shows the configuration of the OTDM receiver [106]. The MLLD-OCR consists of a mode-locked LD with an integrated EAM gate section (EA-MLLD) and a regeneration loop. When the loop gain exceeds unity, only the RF signal near 40 GHz is positively fed back until the output power of the RF amplifier saturates. As a result, a strong

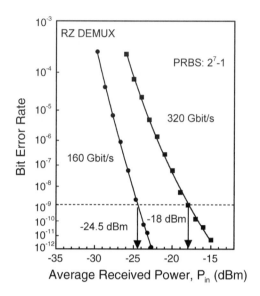

Figure 4.45 Measured BER for DEMUX with the PD-EAM optical gate at 160 and 320 Gbit/s [8]. (Reproduced by permission of © 2003 IET)

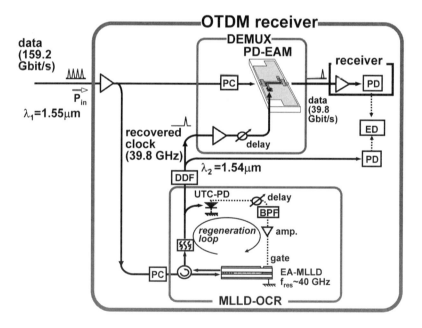

Figure 4.46 Schematic diagram of the OTDM receiver using the PD-EAM and MLLD-OCR [106]. (Reproduced by permission of © 2004 IET)

and stable clock signal can be obtained. The pulse width of the recovered clock after compression by a dispersion decreasing fiber (DDF) was about 1 ps. For the DEMUX experiment, a 159.2-Gbit/s data stream (prepared by optically multiplexing a 39.8-Gbit/s PRBS signal) with the center wavelength of 1.55 µm and the pulse width of 0.5 ps was used as the input signal. The average power of the input data signal and input clock signal for the PD-EAM were 2 and 19 dBm, respectively. Figure 4.47 shows measured BERs for demultiplexed 40-Gbit/s data using either a reference clock or a recovered clock [106]. Although a penalty of 1 to 3 dB arises compared with demultiplexing using the reference clock, error-free operation (BER $\leq 10^{-9}$) for 160- to 40-Gbit/s DEMUX was achieved using the recovered clock from the MLLD-OCR [106]. These results demonstrate that a combination of the PD-EAM and the MLLD-OCR offers a practical OTDM receiver solution for the use in ultrafast optical signal-processing systems.

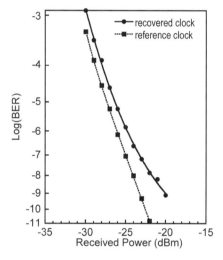

Figure 4.47 BER for 40-Gbit/s data signal demultiplexed from a 160-Gbit/s data stream using the OTDM receiver [106]. (Reproduced by permission of © 2004 IET)

4.4.6.2 Wavelength Conversion

Wavelength conversion is also an important function for transmitters and routers. Although several ultrafast wavelength converters based on all-optical devices have been reported so far [107–109], one of the advantages of an OE-device-based wavelength converter [110] is that there is no interaction between input and output optical signals in principle, because the input optical signal is once converted to electrical signal in the device. Thus, it can be stably operated even when the wavelengths of the signals are very close to each other. The PD-EAM can also be used as an ultrafast wavelength converter having the features mentioned in the previous section.

Figure 4.48 shows the experimental set-up for the 100-Gbit/s wavelength conversion with the PD-EAM [9]. The 100-GHz optical clock signal fed into the EAM was prepared by OTDM of a 12.5-GHz pulse stream generated from a wavelength-tunable ML-FL. The 100-Gbit/s RZ

PRBS optical data signal fed into the UTC-PD was prepared by OTDM of 12.5-Gbit/s optical PRBS data (pattern length: 2^7-1) generated from another ML-FL and a lithium niobate MZM (LN-MZM). The optical MUX for OTDM consists of optical couplers and fiber delay lines. The delay time of the delay lines in the MUX was properly selected to maintain the PRBS (pattern length: 2^7-1) after multiplexing. The wavelengths of the clock (λ_{clock}) and the input data (λ_{in}) signals were changed in the range of 1.545–1.565 and 1.545–1.570 µm (around the C-band), respectively. The 100-Gbit/s output (wavelength-converted) data signal ($\lambda_{out}(=\lambda_{clock})$) from the PD-EAM was fed into a variable optical attenuator (VOA) and then demultiplexed to a 12.5-Gbit/s data signal by a 100-Gbit/s receiver system, which employs another PD-EAM as a DEMUX. The received optical power (P_{in}) used in BER measurement was defined as the average input power to the 100-Gbit/s receiver system. For the wavelength conversion operation of the PD-EAM, input pulse energies to the EAM and the UTC-PD were kept at 10 fJ/pulse and 3 pJ/pulse, respectively. These energies correspond to average power of 0 dBm and 21.76 dBm, respectively. Applied bias voltages of the UTC-PD were kept at -2.0 V. Bias voltages of the EAM were set to an appropriate value for each wavelength of the clock signal [9]. In this experiment, the temperature of the PD-EAM was controlled to be 25 °C.

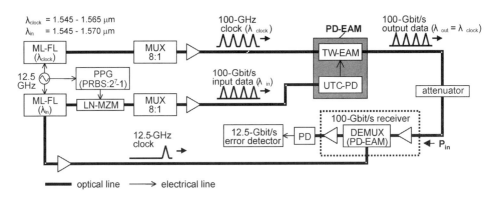

Figure 4.48 Experimental set-up for 100-Gbit/s wavelength conversion with the PD-EAM optical gate. (Reproduced by permission of © 2005 IEEE)

To characterize wavelength conversion performance, the BER was measured for the 100-Gbit/s wavelength conversion while λ_{in} and λ_{out} were varied independently. Figures 4.49(a) and (b) respectively show the BER curves for the wavelength conversions from λ_{in} of 1.545–1.570 µm to λ_{out} of 1.5525 µm, and from λ_{in} of 1.5525 µm to λ_{out} of 1.545–1.565 µm. The BERs measured without conversion are also shown as references. The references in Figure 4.49(b) show different BERs for each wavelength, which is caused by the wavelength dependence of the 100-Gbit/s receiver system. The variation in the sensitivity of the 100-Gbit/s receiver system was less than 1.3 dB in the wavelength range of 1.545–1.565 µm. As shown in Figures 4.49(a) and (b), error-free wavelength conversion at a bit rate of 100 Gbit/s was successfully achieved for all combinations of wavelength [9]. Here, EAM bias voltages were between -1.6 and -2.3 V for λ_{out} of 1.545–1.565 µm. Power penalties, defined as the differences in P_{in} at a BER of 10^{-9} for wavelength conversion and the reference, were less than 2.0 and 2.1 dB for the wavelength range of 25 (λ_{in}) and 20 nm (λ_{out}), respectively. To our knowledge, this is the fastest

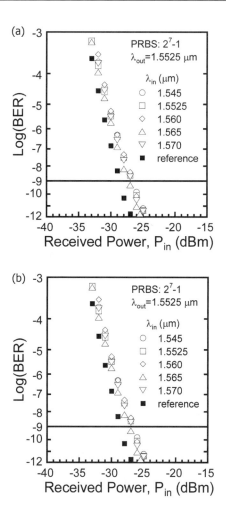

Figure 4.49 BER curves of 100-Gbit/s wavelength conversion for (a) various λ_{in} and (b) various λ_{out} [9]. (Reproduced by permission of © 2005 IEEE)

error-free wavelength conversion ever achieved using an OE-type optical gate. Small penalty fluctuations of 0.6 and 0.4 dB were also confirmed for the wavelength change of input and output data signals, respectively. These small wavelength dependences result from the nearly constant responsivity of the UTC-PD and the extinction ratio of the EAM with a proper bias setting. These results indicate that the PD-EAM is also promising for wavelength conversion in ultrafast optical signal processing.

4.4.6.3 Retiming

In the transmitter, repeater, and router, the 3Rs (retiming, reshaping, and regenerating) are essential functions. Among them, the PD-EAM is promising for 2R operation (retiming and reshaping). Here, we focus on the retiming function of the PD-EAM. As described in 4.4.3, the

PD-EAM can produce a rectangular-like gating waveform. Thus, even when the data pulses for the UTC-PD experience a timing shift against the clock pulses for the EAM within the timing margin corresponding to the width of the rectangular-like gating waveform, the PD-EAM can properly gate the clock pulses to accomplish retiming as shown in Figure 4.31.

Figure 4.50 shows the experimental set-up for the 100-Gbit/s retiming operation of the PD-EAM [111]. A 12.5-GHz optical pulse stream was generated by a ML-FL. The wavelength and FWHM of the initial pulse stream were 1.55 μm and 1.8 ps, respectively. A 100-GHz clock, which was fed into the EAM, was prepared by optically multiplexing the initial pulse stream. The same initial pulse stream was encoded to a PRBS with a pattern length of $2^7 - 1$ by using an LN-MZM. The 100-Gbit/s data, which was fed into the UTC-PD, was prepared by optically multiplexing the 12.5-Gbit/s PRBS. The output data after the retiming by the PD-EAM was then demultiplexed to 12.5-Gbit/s data by a 100-Gbit/s receiver system containing another PD-EAM. To characterize the retiming operation of the PD-EAM, we measured BERs of the demultiplexed 12.5-Gbit/s data. Here, the received power (P_{in}) was again defined as the average optical power of the data signal fed into the 100-Gbit/s receiver system. In addition, we also characterized the retiming margin by shifting the timing (Δt) between clock and input data. Pulse energies of the clock and data before the retiming were 50 fJ/pulse and 2 pJ/pulse, respectively. Bias voltages of the UTC-PD and the EAM were kept at -1.7 and -2.0 V, respectively.

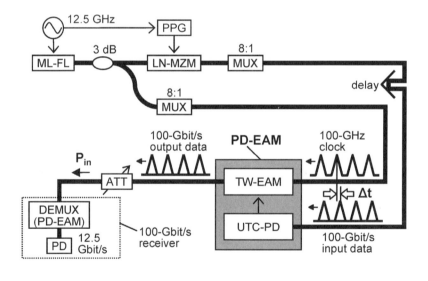

Figure 4.50 Experimental set-up for characterizing 100-Gbit/s retiming operation with the PD-EAM optical gate [111]. (Reproduced by permission of © 2004 IET)

Figure 4.51 shows the measured BER for the 100-Gbit/s retiming at a Δt of 0 ps [111]. For comparison, the BER curve measured by directly demultiplexing the 100-Gbit/s input data (bypassing the PD-EAM for retiming) is also shown. As shown in Figure 4.51, a 100-Gbit/s error-free retiming operation of the PD-EAM was achieved with a receiver sensitivity of -27.5 dBm at a BER of 10^{-9} [111]. In addition, the two curves are very close, which means that the power penalty for the retiming operation by the PD-EAM is very small. In order to estimate the timing margin for the 100-Gbit/s retiming, the BER at various Δt was also measured. Figure 4.52 shows the Δt dependence of power penalty at a constant BER of 10^{-9} [111]. The power penalty rises rapidly when Δt is larger or smaller than $+1$ or -1 ps,

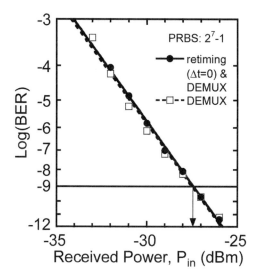

Figure 4.51 BER curves for 100-Gbit/s retiming operation with the PD-EAM optical gate [111]. (Reproduced by permission of © 2004 IET)

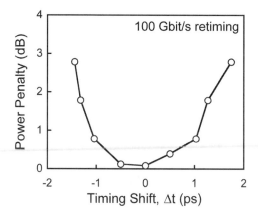

Figure 4.52 Power penalty for various timing shifts at constant BER of 10^{-9} [111]. (Reproduced by permission of © 2004 IET)

respectively. We also found that the timing margin at the power penalty of below 1 dB is 2.2 ps. This is 22 % of a time slot for a 100-Gbit/s data signal. This low power penalty with a wide timing margin indicates that the PD-EAM is a practical device for achieving the 100-Gbit/s retiming function.

4.4.6.4 Optical Sampling

Observation of ultrafast optical signal waveforms is important for developing various optical devices and signal-processing systems. The key component in an instrument for such a purpose is an optical sampling gate that can extract the time-resolved optical intensity of the original waveform. The important criteria for an optical sampling gate are operation speed, signal-to-noise ratio (SNR), device size, operation simplicity, power consumption, and device cost. In this regard, we applied the PD-EAM to the optical waveform observation [11].

As shown in Figure 4.53, the PD-EAM functions as an optical sampling gate when the sampling pulse and data signal are fed into the UTC-PD and EAM, respectively [11]. The base frequency (f_1) of the input data signal and repetition rate (f_2) of the sampling pulses have a slight frequency difference ($\Delta f = f_1 - f_2$). Thus, the trace of the input signal can be sampled and observed using a slower PD and a digital oscilloscope with a time-expanding ratio of $f_1/\Delta f$. Figure 4.54 shows the experimental set-up for optical sampling with the PD-EAM as the sampling gate [11]. The optical sampling system mainly consists of a PD-EAM, a semiconductor-based gain-switched laser diode (GSW-LD), a synthesizer, a low-speed PD, and a digital oscilloscope. The optical sampling pulse stream was generated by the GSW-LD at a repetition rate (f_2) of 9.920 GHz. By compensating for chirp with a dispersion-compensating fiber (DCF) and using a DDF, the pulse was compressed to a width of 1 ps. The optical sampling rate was reduced to 620 MHz by using a frequency divider and an EAM because of the limited external electrical clock rate of the digital oscilloscope. This optical sampling pulse was fed into the UTC-PD, and the input data signal was gated (sampled) at the PD-EAM. After O/E conversion by the low-speed PD, the sampled data signal was fed to the digital oscilloscope together with the synchronized electrical clock. The trigger signal for the digital oscilloscope,

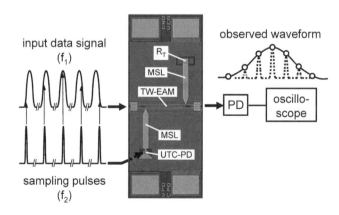

Figure 4.53   Optical sampling using the PD-EAM as a sampling gate [11]. (Reproduced by permission of © 2004 IET)

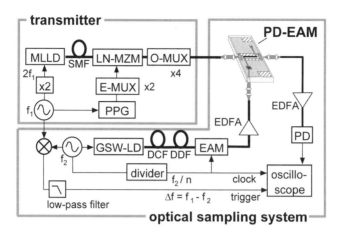

Figure 4.54 Experimental set-up for optical sampling with the PD-EAM optical gate [11]. (Reproduced by permission of © 2004 IET)

which had a Δf of 1 MHz between the f_1 of 9.921 GHz and the f_2 of 9.920 GHz, was generated by a frequency mixer and a low-pass filter. To prepare an input optical data signal, a pulse stream at a repetition rate of twice $f_1(2 \times 9.921$ GHz) was generated from an MLLD and encoded as a PRBS with a pattern length of 2^7-1 by using an LN-MZM. Then, the 19.842-Gbit/s PRBS was optically multiplexed to be a 79.368-Gbit/s data signal. The digital oscilloscope used in this work had a gating-time window of 10 ps. This results in a 22-dB improvement in the SNR of the optically sampled data signal. The pulse energy of the input data signal and optical sampling pulse were kept at 100 fJ/pulse and 7 pJ/pulse, respectively.

Figure 4.55 shows the estimated SNR of the electrically sampled data signal in the digital oscilloscope against V_{EAM} for the PD-EAM [11]. Insets (a) and (b) show signal waveforms of fixed data patterns observed at V_{EAM} of -2.9 and -2.4 V, respectively. The SNR rapidly decreases at a V_{EAM} of less than -3.1 V. At such deep bias voltages compared with the threshold voltage of the EAM (-2.0 V), the effective voltage applied to the EAM becomes very small with a constant optical input to (electrical output from) the UTC-PD due to the nonlinear extinction characteristics of the EAM. Thus, the extinction ratio of the PD-EAM degrades, resulting in a decrease of the SNR. On the other hand, at a V_{EAM} higher than -3.1 V, the SNR increases to more than 10 dB, which is sufficient for the signal waveform observation as shown in the insets. However, at a higher V_{EAM} of -2.4 V, the observed pulse width becomes larger than that at -2.9 V. This is because, when the V_{EAM} is relatively small, the nonlinear extinction in the EAM is not prominent enough to make the gate opening narrower than the width of the electrical pulse generated from the UTC-PD. Thus, we found that a V_{FAM} of -2.9 V is optimum for the optical sampling operation of fast signals. At this bias condition, a clear eye-pattern of 80-Gbit/s RZ-PRBS data signals was successfully observed as shown in Figure 4.56 [11]. Under the same conditions, the gate-opening time obtained from the photoresponse of the PD-EAM using a pump-and-probe technique was found to be 4 ps. The refinement of the PD-EAM, which resulted in a narrower gate opening time of around 2 ps with the same extinction ratio as used in this experiment, implies that optical sampling of 160-Gbit/s-class RZ data signal could be possible with the PD-EAM optical gate.

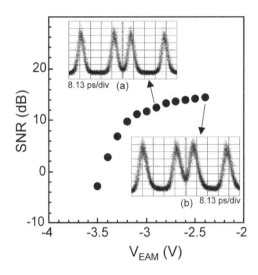

Figure 4.55 SNR against EAM bias voltage. Insets are 80-Gbit/s optical data patterns observed at EAM bias voltages of (a) −2.9 and (b) −2.4 V [11]. (Reproduced by permission of © 2004 IET)

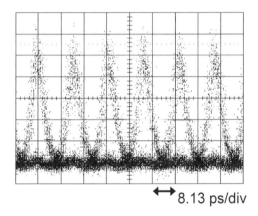

Figure 4.56 Observed eye pattern of 80-Gbit/s PRBS optical data signal at V_{EAM} of −2.9 V [11]. (Reproduced by permission of © 2004 IET)

4.4.6.5 Other Functions

In addition to the functions described above, the PD-EAM can also be used for other functions a good example being as a reshaping function. The optical signal transmitted in optical communications systems is frequently affected by accumulated amplified spontaneous emission (ASE) noise from the optical amplifier, especially at the 'on' (stronger light intensity) level. The degraded SNR of the transmitted signal is usually recovered by a reshaping function in a repeater. Such a function can be achieved cost effectively by using the PD-EAM as a 2R (reshaping and retiming) repeater. As illustrated in Figure 4.57, the EAM exhibits S-shaped nonlinear transmission characteristics with applied voltage. Therefore, if the bias voltage for the EAM and the input power to the UTC-PD are properly adjusted so that the noise components (at 'on' and 'off' levels) lie at the rather flat regions of the transmission characteristic, the variation in the 'on' and 'off' levels will be suppressed. Thus, we can expect the SNR to be improved considerably. This concept has actually been experimentally demonstrated using a discrete-module connection scheme, and the sensitivity of the 10-Gbit/s NRZ signal transmission was improved by 3 dB compared with that without the reshaping function [10]. Replacing those modules with the monolithic PD-EAM optical gate and incorporating the retiming function described in Section 4.4.6.3 would provide a simple 2R (retiming and reshaping) repeater for much higher data rates.

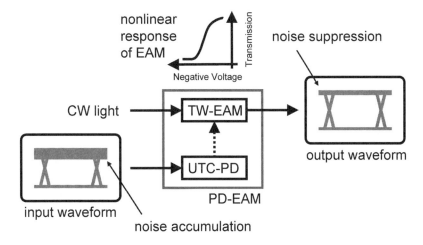

Figure 4.57 Schematic diagram showing how the reshaping operation with a PD-EAM optical gate is accomplished

In addition to reshaping, the PD-EAM can be used for several other functions, such as optical MUX, optical clock recovery, short optical pulse generation, and optical signal inversion. Because an inverter is one of the key functions in electrical logic circuits, further advances in the integration process may lead to a PD-EAM-based optical logic circuit in the future.

4.4.7 Future Work

4.4.7.1 Improving the PD-EAM Optical Gate

Although the PD-EAM optical gate already achieves excellent performance in various capacities, there are still some technical issues to overcome for operations at higher data rates. For improved TW-EAM performance, precise velocity matching between propagating electrical and optical signals is important. Several ideas for accomplishing this have been reported [112, 113], and they should be properly incorporated into the PD-EAM design. The characteristic impedance of the EAM waveguide should be increased a little more to decrease the required photocurrent from the UTC-PD, because there is a trade-off between operating speed and the maximum output current. This can be primarily achieved by making the EAM mesa narrower, which could be achieved by dry etching. The electrical loss in the electrical signal line on the EAM structure also has to be reduced. This can be done by optimizing the epitaxial layer structure of the EAM and the cross-sectional structure of the TW electrode. In addition, the layer structure of the EAM should be reconsidered carefully, because, for very high operation speeds, carrier dynamics, such as carrier escape time from the quantum wells and carrier transit time in the depletion region, will be limiting factors. For improved UTC-PD performance, optimizations among output current, bandwidth, and efficiency are necessary, because there are trade-offs between these parameters. For this purpose, the several guidelines described in Section 4.4.2 would be effective. By combining all of these measures, the operation speed of the PD-EAM can be further improved. The implementation of a waveguide PD structure in the PD-EAM would also be advantageous for constructing a planar integrated circuit with higher complexity. Towards this aim, integrating more than three elemental OE devices, such as a LD and semiconductor-based optical amplifiers in addition to a PD and an EAM [114], is also attractive for realizing higher functionality. This would also be effective for expanding the role of the PD-EAM in ultrafast all-optical signal processing devices.

4.4.7.2 Monolithic PD-MZ Optical Gate

A different approach to improving the device performance is to replace the elemental device that limits the characteristics as a whole. Recently, an InGaAlAs/InAlAs MQW n-i-n MZM has been developed [115]. This optical modulator can achieve both better velocity matching and higher characteristic impedance than can the TW-EAM. In addition, the electrical loss in the electrical signal line is smaller than that of the TW-EAM because the n-type cladding layer of the MZM has a lower resistance than does the p-type cladding layer of the EAM. Furthermore, the MZM functions as a switch with two input and two output ports, which is more advantageous than the simple gating function of the EAM with single input/output ports. Thus, to further improve the performance of the PD-EAM, we have proposed a novel optical gate that monolithically integrates a UTC-PD and InP-based MZM (PD-MZ).

The PD-MZ is schematically shown in Figure 4.58 [116]. It consists of an InP/InGaAs UTC-PD, an InP-based MZM, a terminal resistor, bias capacitors, and thin-film MSLs for interconnections. The chip size is 4.5×1 mm^2. In this device, the MZM is directly driven by electrical output from the UTC-PD, which means, as in the PD-EAM, there is no need for an electrical driver amplifier. The 3-dB bandwidth of the UTC-PD was designed to be 110 GHz.

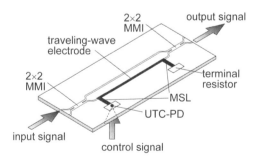

Figure 4.58 Schematic view of a monolithic PD-MZ optical gate [116]. (Reproduced by permission of © 2005 IET)

The MZM consists of two 2×2 multimode interference (MMI) couplers and 2-µm-wide waveguides. The phase-shift region is 3 mm long. The waveguides consist of an n-type contact layer, an n-InP lower cladding layer, an undoped InGaAlAs/InAlAs MQW layer, and an upper cladding layer (which includes an undoped InP layer, a current blocking layer, and an n-type contact layer). TW electrodes are formed on top of the waveguides in the phase shift region. All the epitaxial layers for the UTC-PD and the MZM were successively grown by MOVPE on S.I. InP substrate.

The interaction length between electrical and optical signals in the PD-MZ is typically 15-times longer than in the PD-EAM. Thus, propagation velocity matching between electrical and optical signals becomes more important. In addition, the characteristic impedance (Z_0) of the MZM can be made higher than that for the TW-EAM so that we can reduce the current level required for the UTC-PD. Figure 4.59 shows the electrical/optical propagation velocity ratio (v_e/v_o) and Z_0 of the MZM calculated using an electromagnetic simulator [116]. To obtain single-mode propagation, we set the width of the optical waveguide to 2 µm in this calculation. As shown in Figure 4.59, v_e and Z_0 can be controlled by the width (w) of the TW electrode. As a result, velocity matching is properly accomplished at w of 5 µm with a Z_0 of 35 Ω, which is much higher than the $Z_0(15\Omega)$ of the TW-EAM. The PD-MZ, like the PD-EAM, requires no 50-Ω electrical design. Thus, we employed a w of 5 µm in the fabricated PD-MZ. For impedance matching, the Z_0 of the MSLs and R_T were also designed to be 35Ω.

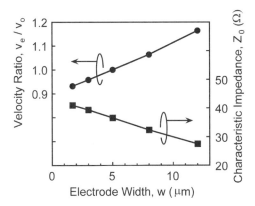

Figure 4.59   Calculated electrical/optical propagation velocity ratio and characteristic impedance of an InP-based MZM [116]. (Reproduced by permission of © 2005 IET)

To characterize the on/off ratio and gate opening times of the fabricated PD-MZ, the pulse photoresponse was measured by the pump-and-probe method. The experimental set-up was the same as that shown in Figure 4.36. The repetition rate, width, and center wavelength of the pump and probe pulses were 52 MHz, 600 fs, and 1540 µm, respectively. Average optical input power to the MZM was −10 dBm. The bias voltage applied to the UTC-PD was −5 V, and the bias voltage applied to the MZM was chosen so that the PD-MZ would function as a transmission-type optical gate. The static extinction ratio for the fabricated MZM was measured as 30.5 dB with a low driving voltage of 1.8 V. Figure 4.60 shows a typical pulse photoresponse curve for the PD-MZ [116]. Here, the average photocurrent of the UTC-PD was

16μA. The gating profile is very smooth and nearly symmetric. In addition, at around 60 ps, which corresponds to the round-trip propagation time from the UTC-PD to terminal resistor through the 3-mm-long TW electrode, there is no reflected signal. This results from smooth electrical signal propagation due to good impedance matching among the MZM, MSLs and terminal resistor. Figure 4.61 shows the on/off ratio and gate opening time (FWHM) of the pulse photoresponse as a function of the average photocurrent of the UTC-PD [116]. Here, to consider the possibility of 100-Gbit/s full-rate operation, the on/off ratio was defined as the ratio between intensities at the peak position and at 10 ps after the peak position. The on/off ratio has a superlinear dependence on the average photocurrent, while the FWHM stays almost constant. The former might be caused by the sinusoidal extinction characteristic of the MZM. As shown in Figure 4.61, a high on/off ratio of more than 10 dB and a small FWHM of about 5 ps were obtained at photocurrents larger than 19 μA. To our knowledge, this is the shortest gate opening time ever achieved using a semiconductor-based MZM. The maximum on/off ratio of 15.7 (11.9 dB) was achieved at the photocurrent of 26 μA.

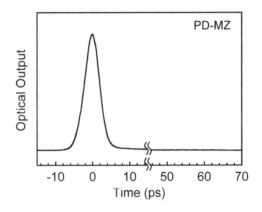

Figure 4.60 Pulse photoresponse curve of a fabricated PD-MZ optical gate [116]. (Reproduced by permission of © 2005 IET)

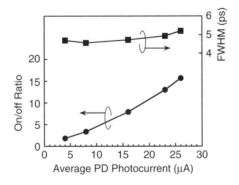

Figure 4.61 On/off ratio and gate opening time (FWHM) of a PD-MZ optical gate [116]. (Reproduced by permission of © 2005 IET)

When the PD-MZ is used as a 1:2 optical DEMUX for example, two demultiplexed signals (one bit shifted from each other) come out from either of the output ports. Thus, it has functionality corresponding to two PD-EAMs. When two half-rate signals, one-bit shifted from each other, are fed to the two input ports of the PD-MZ, it functions as a switched MUX, with which we can select the output port of the multiplexed signal. These results and considerations indicate that the PD-MZ has potential for operation at 100 Gbit/s or more, higher functionality than that of the PD-EAM and so it is a promising candidate for future all-optical signal processing systems.

4.5 Summary and Prospects

The first part of this chapter reviewed the basics of the uni-traveling-carrier photodiode (UTC-PD), such as operation, fabrication, bandwidth, carrier transport, effect of self-induced field, output level, low-bias-voltage operation, efficiency, and reliability. The UTC-PD has a unique mode of operation and is capable of high-speed and high-output current operations. UTC-PDs exhibit a record 3-dB bandwidth of over 300 GHz and a very high peak output current of 184 mA with a 3-dB bandwidth of 65 GHz. As a promising digital application, a UTC-PD photo receiver has also been demonstrated at bit rates of up to 160 Gbit/s with a sufficiently high output voltage to directly drive digital circuit without using electrical post-amplifiers.

For further improving the performance of the UTC-PD and expanding its application area, there are several approaches, many of which have been applied to the conventional pin-PDs. For example, to ultimately increase its operation speed, implementation of the traveling-wave design [68] should be effective [71]. For very-high-output current operations, techniques for managing the heat generation from the devices, which are widely used in electrical power amplifiers, will be necessary, in addition to optimization of the device structure [61]. The use of different material systems [117] is also of interest for widening the wavelength range of the input light signal from infrared to ultraviolet, which would expand the application area of the UTC-PD to various measurement and sensing systems. For this purpose, the built-in field in the absorption layer [1, 18] should be exploited because the benefit of the high minority electron mobility in the absorption layer is best utilized in InGaAs. The operation of the UTC-PD at very low temperatures [118, 119] is also of interest for realizing a high-speed signal interface in future cryogenic electronics [120].

The role of the UTC-PD as a high-output O/E interface device will become more important in future high-bit-rate systems of beyond 100–160 Gbit/s because the capability to drive the electrical circuit directly can take over the intense demands placed on the electrical amplifiers. For this purpose, monolithic integration of a UTC-PD with electron devices is a steady approach to overcoming the restrictions in coaxial connector technology. Novel hybrid approaches for high-speed electrical signal connection involving a wafer-bonding technique [121] are also promising. Another important role of the UTC-PD is as a high-speed and high-output opto-electronic driver. The features of the UTC-PD are indispensable for the direct drive of high-speed E/O devices without electrical amplifiers, and can open-up a new class of opto-electronic integrated device so that the superior performance of each elemental OE device is effectively fused as an integrated ultrafast functional device.

As a promising example of such a novel functional device, the latter part of this chapter described the concept, configuration, design, fabrication, and performance of a monolithic

PD-EAM optical gate that integrates a UTC-PD and a TW-EAM. The PD-EAM exhibits excellent performance, such as 320-Gbit/s-to-10-Gbit/s error-free DEMUX operation, 100-Gbit/s error-free wavelength conversion, and 100-Gbit/s error-free retiming operation. These are the highest operation speeds ever achieved by a semiconductor OE-device-based optical gate. There are many more functions that can be realized using the PD-EAM, and thus, the PD-EAM can be considered as a promising elemental device for ultrafast all-optical signal processing systems.

There are two approaches to further improving the performance of the monolithic PD-EAM optical gate. One is to improve the performance of each element in the PD-EAM. In addition to the UTC-PD, improvements in the TW-EAM are still needed, especially precise velocity matching between propagating electrical and optical signals, an increase in the characteristic impedance, a reduction of electrical loss in the electrical signal line, and proper design of the epitaxial layer structure considering carrier dynamics in the depletion region. Another approach is to employ a different element with better performance. For this purpose, we have proposed a novel PD-MZ [116] optical gate that integrates a UTC-PD and an InP-based MZM [115]. With this configuration, better velocity matching, higher characteristic impedance, and smaller electrical loss in the electrical signal line can be achieved. In addition, it has higher function-ality than the PD-EAM, implying that the PD-MZ is a promising candidate for an integrated optical functional device with higher complexity. These novel OE-based functional devices are expected to be key elements for future ultrafast all-optical signal processing systems.

References

[1] T. Ishibashi, N. Shimizu, S. Kodama, H. Ito, T. Nagatsuma, and T. Furuta, 'Uni-traveling-carrier photodiodes,' *Tech. Dig. Ultrafast Electronics and Optoelectronics* (1997 OSA Spring Topical Meeting), pp. 166–168 (1997).

[2] H. Ito, S. Kodama, Y. Muramoto, T. Furuta, T. Nagatsuma, and T. Ishibashi, 'High-speed and high-output InP-InGaAs unitraveling-carrier photodiodes,' *IEEE J. Selected Topics in Quantum Electron.*, **10**, 709–727 (2004).

[3] K. Sano, K. Murata, T. Otsuji, T. Akeyoshi, N. Shimizu, and E. Sano, 'An 80-Gb/s optoelectronic delayed flip-flop IC using resonant tunneling diodes and uni-traveling-carrier photodiode,' *IEEE J. Solid State Circuits*, **36**, 281–289 (2001).

[4] S. Kodama, T. Ito, N. Watanabe, S. Kondo, H. Takeuchi, H. Ito and T. Ishibashi, '2.3 picoseconds optical gate monolithically integrating photodiode and electroabsorption modulator,' *Electron. Lett.*, **37**, 1185–1186 (2001).

[5] R. Scollo, H.-J. Lohe, J. F. Holzman, F. Robin, H. Jäckel, D. Erni, W. Vogt, and E. Gini, 'Mode-locked laser diode with an ultrafast integrated uni-traveling carrier saturable absorber,' *Optics Lett.*, **30**, 2808–2810 (2005).

[6] N. Kashio, K. Kurishima, K. Sano, M. Ida, N. Watanabe, and S. Yamahata, 'Monolithic integration of InP HBTs and uni-traveling-carrier photodiodes using nonselective regrowth,' *IEEE Trans. Electron Dev.*, **54**, 1651–1657 (2007).

[7] J. W. Raring, L. A. Johansson, E. J. Skogen, M. N. Sysak, H. N. Poulsen, S. P. DenBaars, and L. A. Coldren, '40 Gbit/s photonic receivers integrating UTC photodiodes with high- and low-confinement SOAs using quantum well intermixing and MOCVD regrowth,' *Electron. Lett.*, **42**, 942–943 (2006).

[8] S. Kodama, T. Yoshimatsu and H. Ito, '320-Gbit/s error-free demultiplexing by using an ultrafast optical gate monolithically integrating a photodiode and electroabsorption modulator,' *Electron. Lett.*, **39**, 1269–1270 (2003).

[9] T. Yoshimatsu, S. Kodama, K. Yoshino and H. Ito, '100-Gb/s error-free wavelength conversion with a monolithic optical gate integrating a photodiode and electroabsorption modulator,' *IEEE Photonics Technol. Lett.*, **17**, 2367–2369 (2005).

[10] Y. Kisaka, A. Hirano, M. Yoneyama, and N. Shimizu, 'Simple 2R repeater based on EA modulator directly driven by uni-travelling-carrier photodiode,' *Electron. Lett.*, **35**, 1016–1017 (1999).

[11] S. Kodama, T. Shimizu, T. Yoshimatsu, K. Yoshino, T. Furuta, and H. Ito, 'Ultrafast optical sampling gate monolithically integrating a PD and EAM,' *Electron. Lett.*, **40**, 696–697 (2004).

[12] J. E. Bowers and C. A. Burrus, Jr., 'Ultrawide-band long-wavelength p-i-n photodetectors,' *IEEE J. Lightwave Tech.*, **5**, 1339–1350 (1987).

[13] K. Hagimoto, Y. Miyamoto, T. Kataoka, H. Ichino, and O. Nakajima, 'Twenty-Gbit/s signal transmission using simple high-sensitivity optical receiver,' *Tech. Dig. 16th Optical Fiber Communication Conference*, p. 48, (1992).

[14] Y. Miyamoto, M. Yoneyama, T. Otsuji, K. Yonenaga, and N. Shimizu, '40-Gbit/s TDM transmission technologies based on high-speed ICs,' *IEEE Journal of Solid-State Circuits*, **34**, 1246–1253 (1999).

[15] T. Furuta, H. Ito, and T. Ishibashi, 'Photocurrent dynamics of uni-traveling-carrier and conventional pin-photodiodes,' *Inst. Phys. Conference Ser.*, No. 166, pp. 419–422 (2000).

[16] M. Yoneyama, Y. Miyamoto, T. Otsuji, H. Toba, Y. Yamane, T. Ishibashi, and H. Miyazawa, 'Fully electrical 40-Gb/s TDM system prototype based on InP HEMT digital IC technologies,' *IEEE J. Lightwave Technol.*, **18**, 34–43 (2000).

[17] H. Ito, T. Furuta, F. Nakajima, K. Yoshino, and T. Ishibashi, 'Photonic generation of continuous THz wave using uni-traveling-carrier photodiode,' *IEEE J. Lightwave Technol.*, **23**, 4016–4021 (2005).

[18] T. Ishibashi, S. Kodama, N. Shimizu, and T. Furuta, 'High-speed response of uni-traveling carrier photodiodes,' *Jpn J. Appl. Phys.*, **36**, 6263–6268 (1997).

[19] T. Ishibashi, T. Furuta, H. Fushimi, S. Kodama, H. Ito, T. Nagatsuma, N. Shimizu, and Y. Miyamoto, 'InP/InGaAs uni-traveling-carrier photodiodes,' *IEICE Trans. Electron.*, **E83-C**, 938–949 (2000).

[20] T. Ishibashi, T. Furuta, H. Fushimi, and H. Ito, 'Photoresponse characteristics of uni-traveling-carrier photodiodes,' *Proceedings SPIE*, **4283**, 469–479 (2001).

[21] T. Ishibashi, 'Nonequilibrium electron transport in HBTs,' *IEEE Trans. Electron Dev.*, **48**, 2595–2605 (2001).

[22] J. L. Moll and I. M. Ross, 'The dependence of transistor parameters on the distribution of base layer resistivity,' *Proceedings IRE*, **44**, 72–78 (1956).

[23] H. Kroemer, 'Heterostructure bipolar transistors and integrated circuits,' *Proceedings IEEE*, **70**, 13–25 (1982).

[24] R. N. Nottenburg, J.-C. Bischoff, M. B. Panish, and H. Temkin, 'High-speed InGaAs(P)/InP double-heterostructure bipolar transistors,' *IEEE Electron Dev. Lett.*, **8**, 282–284 (1987).

[25] O. Sugiura, A. G. Dentai, C. H. Joyner, S. Chandrasekhar, and J. C. Campbell, 'High-current-gain InGaAs/InP double-heterojunction bipolar transistors grown by metal organic vapor phase epitaxy,' *IEEE Electron Dev. Lett.*, **9**, 253–255 (1988).

[26] H. Hafizi, T. Liu, D. B. Rensch, and W. E. Stanchina, 'Effects of collector doping on DC and RF performance of AlInAs/GaInAs/InP double heterojunction bipolar transistors,' *IEEE Trans. Electron Dev.*, **40**, 2122–2123 (1993).

[27] A. Feygenson, D. Ritter, R. A. Hamm, P. R. Smith, R. K. Montgomery, R. D. Yadvish, H. Temkin, and M. B. Panish, 'InGaAs/InP composite collector heterostructure bipolar transistors,' *Electron. Lett.*, **28**, 607–609 (1992).

[28] T. Ishibashi and Y. Yamauchi, 'A possible near-ballistic collection in an AlGaAs/GaAs HBT with a modified collector structure,' *IEEE Trans. Electron Dev.*, **35**, 401–404, (1988).

[29] T. Ishibashi, 'High speed heterostructure devices,' *Semiconductors and Semimetals*, Volume 41, Chapter 5, Academic Press, San Diego (1994), pp. 332–337.

[30] E. S. Harmon, M. L. Lovejoy, M. R. Melloch, M. S. Lundstrom, D. Ritter, and R. A. Hamm, 'Minority-carrier mobility enhancement in p+ InGaAs lattice matched to InP,' *Appl. Phys. Lett.*, **63**, 636–638 (1993).

[31] D.-H. Jun, I.-H. Kang, K.-H. Oh, J.-S., Lee, and J.-I. Song, 'Speed performance comparison of InP/InGaAs UTC-photodiodes utilizing compositionally graded and exponentially doped photo-absorption layers,' *Proceedings Int. Conference Indium Phosphide and Related Materials*, pp. 647–650 (2002).

[32] K. J. Williams, R. D. Esman, and M. Dagenais, 'Effects of high space-charge fields on the response of microwave photodetectors,' *IEEE Photonics Technol. Lett.*, **6**, 639–641 (1994).

[33] B. R. Nag, S. R. Ahmed, and M. D. Roy, 'Electron velocity in short samples of $Ga_{0.47}In_{0.53}As$ at 300 K,' *IEEE Trans. Electron Dev.*, **33**, 788–791 (1986).

[34] K. Brennan, 'Theory of the steady-state hole drift velocity in InGaAs,' *Appl. Phys. Lett.*, **51**, 995–997 (1987).

[35] B. R. Nag and S. R. Ahmed, 'Threshold field and peak-valley velocity ratio in short samples of InP at 300 K,' *IEEE Trans. Electron Dev.*, **34**, 953–956 (1987).

[36] C. T. Kirk, Jr., 'A theory of transistor cutoff frequency (f_T) falloff at high current densities,' *IRE Trans. Electron Dev.*, **9**, 164–174 (1962).

[37] N. Shimizu, N. Watanabe, T. Furuta, and T. Ishibashi, 'Improved response of uni-traveling-carrier photodiodes by carrier injection,' *Jpn J. Appl. Phys.*, **37**, 1424–1426 (1997).

[38] T. Nagatsuma, M. Yaita, M. Shinagawa, K. Kato, A. Kozen, K. Iwatsuki, and K. Suzuki, 'Electro-optic character-ization of ultrafast photodetectors using adiabatically compressed soliton pulses,' *Electron. Lett.*, **30**, 814–816 (1994).

[39] H. Ito, T. Furuta, S. Kodama, and T. Ishibashi, 'InP/InGaAs uni-travelling-carrier photodiode with a 310 GHz bandwidth,' *Electron. Lett.*, **36**, 1809–1810 (2000).

[40] K. Kato, S. Hata, K. Kawano, and A. Kozen, 'Design of ultrawide-band, high-sensitivity p-i-n photodetectors,' *IEICE Trans. Electron.*,. **E76-C**, 214–221 (1993).

[41] P. Hill, J. Schlafer, W. Powazinik, M. Urban, E. Eichen, and R. Olshansky, 'Measurement of hole velocity in n-type InGaAs,' *Appl. Phys. Lett.*, **50**, 1260–1262 (1987).

[42] T. H. Windhorn, L. W. Cook, M. A. Haase, and G. E. Stillman, 'Electron transport in InP at high electric fields,' *Appl. Phys. Lett.*, vol. 42, 725–727 (1983).

[43] Y. K. Chen, A. F. J. Levi, R. N. Nottenburg, P. H. Beton, and M. B. Panish, 'High-frequency study of nonequilibrium transport in heterostructure bipolar transistors,' *Appl. Phys. Lett.*, **55**, 1789–1791 (1989).

[44] K. Kurishima, H. Nakajima, Y. K. Fukai, Y. Matsuoka, and T. Ishibashi, 'Electron velocity overshoot effect in collector depletion layers of InP/InGaAs heterojunction bipolar transistors,' *Jpn J. Appl. Phys.*, **31**, L768–L770 (1992).

[45] R. L. Tebbenham and D. Walsh, 'Velocity/field characteristics of n-type indium phosphide at 110 and 330 K,' *Electron. Lett.*, **11**, 96–97 (1975).

[46] T. J. Malonney and J. Frey, 'Transient and steady-state electron transport properties of GaAs and InP,' *J. Appl. Phys.*, **48**, 781–787 (1977).

[47] W. T. Masselink and T. F. Kuech, 'Velocity-field characteristics of electrons in doped GaAs,' *J. Electronic Mater.*, **18**, 579–584 (1989).

[48] J. H. Marsh, P. A. Houston, and P. N. Robson, 'Compositional dependence of the mobility, peak velocity and threshold field in $In_{1-x}Ga_xAs_yP_{1-y}$,' *Inst. Phys. Conference Ser.*, No. 56, pp. 621–630 (1981).

[49] J. C. Aquino, T. Kawashima, and M. Tokuda, 'Evaluation of anechoic chamber characteristics using optically driven equipment under test,' *IEEE Int. Symp. on EMC*, pp. 231–233 (1999).

[50] H. Ito, T. Furuta, and T. Ishibashi, 'InP/InGaAs uni-traveling-carrier photodiodes,' *IEICE Trans. Electron.*, **E84-C**, 1448–1454 (2001).

[51] H.-G. Bach, 'Ultrafast waveguide-integrated pin-photodiodes and photonic mixers from GHz to THz range,' *Proceedings 33rd European Conference and Exhibition on Optical Communication*, Volume 5, pp. 247–250 (2007).

[52] H. Ito, T. Nagatsuma, A. Hirata, T. Minotani, A. Sasaki, Y. Hirota, and T. Ishibashi, 'High-power photonic millimetre-wave generation at 100 GHz using matching-circuit, integrated uni-travelling-carrier photodiodes,' *IEE Proceedings Optoelectron.*, **150**, 138–142 (2003).

[53] W.-Y. Hwang, D. L. Miller, Y. K. Chen, and D. A. Humphrey, 'Carbon doping of InGaAs in solid-source molecular beam epitaxy using carbon tetrabromide,' *J. Vac. Sci. Technol.*, **B12**, 1193–1196 (1994).

[54] Y. Takeda, 'Low-field transport calculations,' Chapter 9 in *GaInAsP Alloy Semiconductors* (T. P. Pearsall, ed.), John Wiley & Sons, Inc., New York (1982), p. 226.

[55] K. Y. Cheng, A. Y. Cho, S. B. Christman, T. P. Pearsall, and J. E. Rowe, 'Measurement of the Γ–L separation in $Ga_{0.47}In_{0.53}As$ by ultraviolet photoemission,' *Appl. Phys. Lett.*, **40**, 423–425 (1982).

[56] H. Ito, H. Fushimi, Y. Muramoto, T. Furuta and T. Ishibashi, 'High-power photonic microwave generation at K- and K_a-bands using a uni-traveling-carrier photodiode,' *IEEE J. Lightwave Tech.*, **20**, 1500–1505 (2002).

[57] N. Duan, X. Wang, N. Li, H.-D. Liu, and J. C. Campbell, 'Thermal analysis of high-power InGaAs-InP photodiodes,' *IEEE J. Quantum Electron.*, **42**, 1255–1258, (2006).

[58] N. Shimizu, N. Watanabe, T. Furuta, and T. Ishibashi, 'InP-InGaAs uni-traveling-carrier photodiode with improved 3-dB bandwidth of over 150 GHz,' *IEEE Photonics Technol. Lett.*, **10**, 412–414 (1998).

[59] N. Li, X. Li, S. Demiguel, X. Zheng, J. C. Campbell, D. A. Tulchinsky, K. J. Williams, T. D. Isshiki, G. S. Kinsey, and R. Sudharsanan, 'High-saturation-current charge-compensated InGaAs-InP uni-traveling-carrier photodiode,' *IEEE Photonics Technol. Lett.*, **16**, 864–866 (2004).

[60] J.-W. Shi, Y.-S. Wu, C.-Y. Wu, P.-H. Chiu, and C.-C. Hong, 'High-speed, high-responsivity, and high-power per-formance of near-ballistic uni-traveling-carrier photodiode at 1.55-μm wavelength,' *IEEE Photonics Technol. Lett.*, **17**, 1929–1931 (2005).

[61] K. J. Williams, 'Comparisons between dual-depletion-region and uni-travelling-carrier p-i-n photodetectors,' *IEE Proceedings Optoelectron.*, **149**, 131–137 (2002).

[62] O. Madelung, M. Schultz, and H. Weiss, 'Numerical data in science and technology,' *Landolt-Boernstein*, 17, pp. 571, 619, Springer-Verlag, Berlin (1982).

[63] T. Yasui, T. Furuta, T. Ishibashi, and H. Ito, 'Comparison of power dissipation tolerance of InP/InGaAs UTC-PDs and pin-PDs,' *IEICE Trans. Electron.*, **E86-C**, 864–866 (2003).

[64] M. Yuda, K. Kato, R. Iga, and M. Mitsuhara, 'High-input-power-allowable uni-travelling-carrier waveguide photodiodes with semi-insulating-InP buried structure,' *Electron. Lett.*, 35, 1377–1379 (1999).

[65] D. A. Tulchinsky, X. Li, N. Li, S. Demiguel, J. C. Campbell, and K. J. Williams, 'High-saturation current wide-bandwidth photodetectors,' *IEEE J. Selected Topics in Quantum Electron.*, 10, 702–708 (2004).

[66] J. E. Bowers and C. A. Burrus, 'High-speed zero-bias waveguide photodetectors,' *Electron. Lett.*, 22, 905–906 (1986).

[67] A. Umbach, D. Trommer, A. Siefke, and G. Unterbörsch, '50 GHz operation of waveguide integrated photodiode at 1.55 μm,' *Proceedings 21th European Conference on Optical Communication*, pp. 1075–1078 (1995).

[68] K. S. Giboney, R. L. Nagarajan, T. E. Reynolds, S. T. Allen, R. P. Mirin, M. J. W. Rodwell, and J. E. Bowers, 'Travelling-wave photodetectors with 172-GHz bandwidth and 76-GHz bandwidth-efficiency product,' *IEEE Photonics Technol. Lett.*, 7, 412–414 (1995).

[69] L. Y. Lin, M. C. Wu, T. Itoh, T. A. Vang, R. E. Muller, D. L. Sivco, and A. Y. Cho, 'Velocity-matched distributed photodetectors with high-saturation power and large bandwidth,' *IEEE Photon. Technol. Lett.*, 8, 1376–1378 (1996).

[70] Y. Muramoto, K. Kato, M. Mitsuhara, O. Nakajima, Y. Matsuoka, N. Shimizu, and T. Ishibashi, 'High-output voltage, high speed, high efficiency uni-traveling-carrier waveguide photodiode,' *Electron. Lett.*, 34, 122–123 (1998).

[71] Y. Hirota, T. Ishibashi, and H. Ito, '1.55-μm wavelength periodic traveling-wave photodetector fabricated using unitraveling-carrier photodiode structures,' *J. Lightwave Technol.*, 19, 1751–1758 (2001).

[72] M. Achouche, V. Magnin, J. Harari, F. Lelarge, E. Derouin, C. Jany, D. Carpentier, F. Blache, and D. Decoster, 'High performance evanescent edge coupled waveguide unitraveling-carrier photodiodes for >40-Gb/s optical receivers,' *IEEE Photonics Technol. Lett.*, 16, 584–586 (2004).

[73] H. Fukano, Y. Muramoto, K. Takahata, and Y. Matsuoka, 'High-efficiency edge-illuminated uni-traveling-carrier-structure refracting-facet photodiode,' *Electron. Lett.*, 35, 1664–1665 (1999).

[74] Y. Muramoto, H. Fukano, and T. Furuta, 'A polarization-independent refracting-facet uni-traveling-carrier photodiode with high efficiency and large bandwidth,' *IEEE J. Lightwave Tech.*, 24, 3830–3834 (2006).

[75] H. Ito, T. Furuta, S. Kodama, and T. Ishibashi, 'High-efficiency unitraveling-carrier photodiode with an integrated total-reflection mirror,' *IEEE J. Lightwave Technol.*, 18, 384–387 (2000).

[76] K. Kato, S. Hata, A. Kozen, J. Yoshida, and K. Kawano, 'High-efficiency waveguide InGaAs pin photodiode with bandwidth of over 40 GHz,' *IEEE Photonics Technol. Lett.*, 3, 473–474 (1991).

[77] X. Li, N. Li, X. Zheng, S. Demiguel, J. C. Campbell, D. A. Tulchinsky, and K. J. Williams, 'High-saturation-current InP-InGaAs photodiode with partially depleted absorber,' *IEEE Photonics Technol. Lett.*, 15, 1276–1278 (2003).

[78] Y. Muramoto and T. Ishibashi, 'InP/InGaAs pin photodiode structure maximizing bandwidth and efficiency,' *Electron. Lett.*, 39, 1749–1750 (1992).

[79] D.-H. Jun, J.-H. Jang, I. Adesida, and J.-I. Song, 'Improved efficiency-bandwidth product of modified uni-traveling carrier photodiode structures using an undoped photo-absorption layer,' *Jpn J. Appl. Phys.*, 45, 3475–3478 (2006).

[80] F. J. Effenberger and A. M. Joshi, 'Ultrafast dual-depletion region, InGaAs/InP p-i-n detector,' *IEEE J. Lightwave Tech.*, 14, 1859–1864 (1996).

[81] T. Furuta, H. Fushimi, T. Yasui, Y. Muramoto, H. Kamioka, H. Mawatari, H. Fukano, T. Ishibashi, and H. Ito, 'Reliability study on uni-traveling-carrier photodiode for a 40 Gbit/s optical transmission systems,' *Electron. Lett.*, 38, 332–334 (2002).

[82] H. Ito, T. Ito, Y. Muramoto, T. Furuta, and T. Ishibashi, 'Rectangular waveguide output uni-traveling-carrier photodiode module for high-power photonic millimeter-wave generation in the F-band,' *IEEE J. Lightwave Tech.*, 21, 3456–3462 (2003).

[83] Y. Muramoto, Y. Hirota, K. Yoshino, H. Ito and T. Ishibashi, 'A uni-travelling-carrier photodiode module with a bandwidth of 80 GHz,' *Electron. Lett.*, 39, 1851–1852 (2003).

[84] Y. Muramoto, K. Yoshino, S. Kodama, Y. Hirota, H. Ito, and T. Ishibashi, '100- and 160-Gbit/s operations of uni-traveling-carrier photodiode module,' *Electron. Lett.*, **40**, 378–380 (2004).

[85] S. Kimura, Y. Imai, Y. Umeda, and T. Enoki, 'Loss-compensated distributed baseband amplifier ICs for optical transmission systems,' *IEEE Trans. Microwave Theory and Tech.*, **44**, 1688–1693 (1996).

[86] S. Matsuda, T. Takahashi, and K. Joshin, 'An over-110-GHz InP HEMT flip-chip distributed baseband amplifier with inverted microstrip line structure for optical transmission system,' *IEEE J. Solid-State Circuits*, **38**, 1479–1484 (2003).

[87] N. Shimizu, K. Murata, A. Hirano, Y. Miyamoto, H. Kitabayashi, Y. Umeda, T. Akeyoshi, T. Furuta, and N. Watanabe, '40 Gbit/s monolithic digital OEIC composed of unitraveling-carrier photodiode and InP HEMTs,' *Electron. Lett.*, **36**, 1220–1222 (2000).

[88] K. Murata, K. Sano, T. Akeyoshi, N. Shimizu, E. Sano, M. Yamamoto, and T. Ishibashi, 'Optoelectronic clock recovery circuit using a resonant tunneling diode and uni-travelling-carrier photodiode,' *Electron. Lett.*, **34**, 1424–1425 (1998).

[89] K. Sano, K. Murata, H. Matsuzaki, H. Kitabayashi, T. Akeyoshi, H. Ito, T. Enoki, and H. Sugahara, '75-GHz optical clock device-by-two OEIC using InP HEMT and uni-traveling-carrier photodiode,' *Extended Abstract of the International. Conference on Solid State Devices and Materials*, pp. 902–903 (2003).

[90] K. Sano, K. Murata, T. Akeyoshi, N. Shimizu, T. Otsuji, M. Yamamoto, T. Ishibashi, and E. Sano, 'An ultra-fast optoelectronic circuit using resonant tunneling diodes and a uni-traveling-carrier photodiode,' *Electron. Lett.*, **34**, 215–217 (1998).

[91] K. Sano, K. Murata, T. Otsiji, T. Akeyoshi, N. Shimizu, M. Yamamoto, T. Ishibashi, and E. Sano, 'Ultra-fast optoelectronic decision circuit using resonant tunneling diodes and a uni-traveling-carrier photodiode,' *IEICE Trans. Electron.*, **E82-C**, 1638–1646 (1999).

[92] M. Yoneyama, Y. Miyamoto, K. Hagimoto, N. Shimizu, T. Ishibashi, and K. Wakita, '40 Gbit/s optical gate using optical modulator directly driven by uni-travelling-carrier photodiode,' *Electron. Lett.*, **34**, 1007–1009 (1998).

[93] H. H. Liao, X. B. Mei, K. K. Loi, C. W. Tu, P. M. Asbeck, and W. S. C. Chang, 'Microwave structures for traveling-wave MQW electro-absorption modulators for wide band 1.3 μm photonic links,' *Proceedings SPIE*, 3006, pp. 291–300 (1997).

[94] K. Kawano, M. Kohtoku, M. Ueki, T. Ito, S. Kondoh, Y. Noguchi, and Y. Hasumi, 'Polarisation-insensitive traveling-wave electrode electroabsorption (TW-EA) modulator with bandwidth over 50GHz and driving voltage less than 2V,' *Electron. Lett.*, **33**, 1580–1581 (1997).

[95] K. C. Gupta, R. Garg, I. Bahl, and P. Bharita, *Microstrip lines and Solitons*, second edition, Artech House, Boston (1996), p. 10.

[96] M. Suzuki, H. Tanaka, N. Edagawa, K. Utaka, and Y. Matsushima, 'Transform-limited optical pulse generation up to 20-GHz repetition rate by a sinusoidally driven InGaAsP electroabsorption modulator,' *IEEE J. Lightwave Technol.*, **11**, 468–473 (1993).

[97] K. Wakita, K. Yoshino, A. Hirano, S. Kondo, and Y. Noguchi, '4–5 ps optical pulse generation with 40GHz train from low driving-voltage modulator modules,' *Proceedings 10th International Conference on Indium Phosphide and Related Materials*, pp. 679–682 (1998).

[98] K. Yoshino, T. Yoshimatsu, S. Kodama, and H. Ito, 'Three-optical-port module for monolithic optical gate integrated with photodiode and electroabsorption modulator,' *IEEE Trans. Advanced Packaging*, **29**, 766–769 (2006).

[99] K. Yoshino, T. Takeshita, I. Kotaka, S. Kondo, Y. Noguchi, R. Iga, and K. Wakita, 'Compact and stable electroabsorption optical modulator modules,' *J. Lightwave Technol.*, **17**, 1700–1707 (1999).

[100] S. Kodama, T. Yoshimatsu and H. Ito, '500Gbit/s optical gate monolithically integrating a photodiode and electroabsorption modulator,' *Electron. Lett.*, **40**, 555–557 (2004).

[101] B. Mikkelsen, G. Raybon, R.-J. Essiambre, A. J. Stentz, T. N. Nielsen, D. W. Peckham, L. Hsu, L. Gruner-Nielsen, K. Dreyer, and J. E. Johnson, '320-Gb/s single-channel pseudolinear transmission over 200 km of nonzero-dispersion fiber,' *IEEE Photonics Technol. Lett.*, **12**, 1400–1402 (2000).

[102] H.-F. Chou, Y.-J. Chiu, W. Wang, J. E. Bowers, and D. J. Blumenthal, 'Compact 160-Gb/s demultiplexer using a single-stage electrically gated electroabsorption modulator,' *IEEE Photonics Technol. Lett.*, **15**, 1458–1460 (2003).

[103] T. Ohno, K. Sato, T. Shimizu, T. Furuta and H. Ito, 'Recovery of 40 GHz optical clock from 160Gbit/s data using regeneratively modelocked semiconductor laser,' *Electron. Lett.*, **39**, 453–455 (2003).

[104] T. Ohno, K. Sato, R. Iga, Y. Kondo, K. Yoshino, T. Furuta and H. Ito, 'Recovery of 80 GHz optical clock from 160 Gbit/s data using regeneratively mode locked laser diode,' *Electron. Lett.*, **39**,1398–1400 (2003).

[105] T. Ohno, K. Sato, R. Iga, Y. Kondo, T. Ito, T. Furuta, K. Yoshino, and H. Ito, 'Recovery of a 160-GHz optical clock from 160 Gbit/s data stream using mode locked laser diode,' *Electron. Lett.*, **40**, 265–267 (2004).

[106] T. Ohno, S. Kodama, T. Yoshimatsu, K. Yoshino, and H. Ito, '160-Gbit/s OTDM receiver consisting of a PD-EAM optical-gate and an MLLD-based optical clock recovery device,' *Electron. Lett.*, **40**, 1285–1287 (2004).

[107] S. Nakamura, Y. Ueno, and K. Tajima, '168-Gb/s all-optical wavelength conversion with a symmetric-Mach–Zehnder-type switch,' *IEEE Photon. Technol. Lett.*, **13**, 1091–1093 (2001).

[108] C. Schubert, R. Ludwig, S. Watanabe, F. Futami, C. Schmidt, J. Berger, C. Boerner, S. Ferber, and H. G. Weber, '160 Gbit/s wavelength converter with 3R-regenerating capability,' *Electron. Lett.*, **38**, 903–904 (2002).

[109] K. Nishimura, R. Inohara, M. Usami, and S. Akiba, 'All-optical wavelength conversion by electroabsorption modulator,' *IEEE J. Select. Topics Quantum Electron.*, **11**, 278–284 (2005).

[110] V. A. Sabnis, H. V. Demir, O. Fidaner, J. S. Harris, Jr., D. A. B. Miller, J.-F. Zheng, N. Li, T.-C. Wu, H.-T. Chen, and Y.-M. Houng, 'Optically controlled electroabsorption modulators for unconstrained wavelength conversion,' *Appl. Phys. Lett.*, **84**, 469–471 (2004).

[111] T. Yoshimatsu, S. Kodama, K. Yoshino and H. Ito, '100-Gbit/s error-free retiming operation of monolithic optical gate integrating with photodiode and electroabsorption modulator,' *Electron. Lett.*, **40**, 626–628 (2004).

[112] G. L. Li, C. K. Sun, S. A. Pappert, W. X. Chen, and P. K. L. Yu, 'Ultrahigh-speed traveling-wave electroabsorption modulator – design and analysis,' *IEEE Trans. Microwave Theory and Tech.*, **47**, 1177–1183 (1999).

[113] S. Irmscher, R. Lewen, and U. Eriksson, 'Influence of electrode width on high-speed performance of traveling-wave electro-absorption modulators,' *Proceedings International Conferenceon on Indium Phosphide and Related Materials*, pp. 436–439 (2001).

[114] M. N. Sysak, J. S. Barton, L. A. Johansson, J. W. Raring, E. J. Skogen, M. L. Masanovic, D. J. Blumenthal, and L. A. Coldren, 'Single-chip wavelength conversion using a photocurrent-driven EAM integrated with a widely tunable sampled-gating DBR laser,' *IEEE Photonics Technol. Lett.*, **16**, 2093–2095 (2004).

[115] K. Tsuzuki, T. Ishibashi, T. Ito, S. Oku, Y. Shibata, R. Iga, Y. Kondo, and Y. Tohmori, '40 Gbit/s n-i-n InP Mach-Zehnder modulator with a π voltage of 2.2 V,' *Electron. Lett.*, **39**, 1464–1466 (2003).

[116] T. Yoshimatsu, S. Kodama, and H. Ito, 'InP-based ultrafast optical gate monolithically integrating uni-travelling-carrier photodiode and Mach–Zehnder modulator,' *Electron. Lett.*, **41**, 1243–1244 (2005).

[117] J.-W. Shi, Y.-T. Li, C.-L. Pan, M. L. Lin, Y. S. Wu, W. S. Liu, and J.-I. Chyi, 'Bandwidth enhancement phenomenon of a high-speed GaAs-AlGaAs based unitraveling carrier photodiode with an optimally designed absorption layer at an 830 nm wavelength,' *Appl. Phys. Lett.*, **89**, 053512-1-3 (2006).

[118] K. Yoshino, Y. Muramoto, T. Furuta, and H. Ito, 'High-speed uni-travelling-carrier photodiode module for ultra-low-temperature operation,' *Electron. Lett.*, **41**, 1030–1031 (2005).

[119] H. Ito, T. Furuta, S. Kodama, K. Yoshino, T. Nagatsuma, and Z. Wang, '10-Gbit/s operation of a uni-travelling-carrier photodiode module at 2.6K,' *Electron. Lett.*, **44**, 149–151 (2008).

[120] B. V. Zeghbroeck, 'Optical data communication between Josephson-junction circuits and room-temperature electronics,' *IEEE Transactions on Applied Superconductivity*, **3**, 2881–2884 (1993).

[121] Y. Royter, T. Furuta, S. Kodama, N. Sahri, T. Nagatsuma, and T. Ishibashi, 'Integrated packaging of over 100 GHz bandwidth uni-travelling-carrier photodiodes,' *IEEE Electron Dev. Lett.*, **21**, 158–160 (2000).

5

Intersub-band Transition All-Optical Gate Switches

Nobuo Suzuki, Ryoichi Akimoto, Hiroshi Ishikawa
and Hidemi Tsuchida

5.1 Operation Principle

The intersub-band transitions (ISBT) in semiconductor quantum wells (QWs) [1] can be utilized for ultrafast (>100 Gb/s) all-optical gate switches [2]. A conduction band diagram of a quantum well is illustrated in Figure 5.1. We assume that a high density of electrons locates in the lower sub-band. Without a strong control optical pulse, weak signal light resonant to the ISBT is absorbed. When about half of electrons are excited by a strong control optical pulse, absorption is saturated, and hence the signal light is transmitted. The intersub-band carrier relaxation process is dominated by LO-phonon scattering. Since the intersub-band LO-phonon scattering time is of the order of 0.1–1 ps [3], pattern-effect free, all-optical gate operation at a data rate above 100 Gb/s is achievable with the saturation of the intersub-band absorption. In order to achieve near-infrared absorption, thin QWs (several monolayers (MLs)) with a large band offset are required. Until now, transitions at $\lambda \sim 1.55\,\mu m$ have been achieved only in GaN-based multiple quantum wells (MQWs) [4–8], II–VI MQWs [9, 10] and InGaAs-based coupled quantum wells (CQWs) [11–14]. In the following sections, fundamental properties of the ISBT are explained. Then, the present status of GaN-based, II–VI-based, and InGaAs-based ISBT gate switches are summarized in Sections 5.2, 5.3, and 5.4, respectively. Recently, a novel cross-phase modulation (XPM) switch utilizing the refractive index change due to ISBT in InGaAs/AlAsSb CQWs has been proposed [15]. The mechanism and applications of the XPM switches are discussed in Section 5.5.

Ultrafast All-Optical Signal Processing Devices   Edited by Hiroshi Ishikawa
© 2008 John Wiley & Sons, Ltd

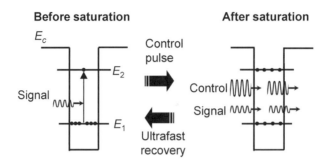

Figure 5.1   Concept of ultrafast all-optical gate switches utilizing the intersub-band transition (ISBT) in semiconductor quantum wells

5.1.1 Transition Wavelength

The intersub-band transition is a transition between quantized energy levels in a quantum well. For simplicity, we first assume an infinite quantum well with a width of L_w. The energy level of the ith sub-band is given by [1]:

$$E_i = i^2 \frac{\hbar^2 \pi^2}{2m^* L_w} \tag{5.1}$$

where $\hbar$ is the Planck constant divided by 2π and m^* is the effective mass. The energy separation between the first and the second sub-bands is inversely proportional to L_w. Therefore, we can tune the transition wavelength range from THz to near-infrared by utilizing the same material system. In an actual well with a finite barrier height, however, the number of quantized states decreases with reducing L_w. In order to achieve near-infrared transition, a thin QW (several MLs) with a large band offset is required.

Figure 5.2 shows conduction band diagrams of a GaN/AlN QW, a (CdS/ZnSe)/BeTe QW and an InGaAs/AlAs/AlAsSb CQW, where transitions at $\lambda \sim 1.55\,\mu m$ have been achieved. GaN/Al(Ga)N QWs have strong built-in fields ($\sim MV/cm$) due to the piezoelectric effect and spontaneous polarization [16]. The field in barriers raises the effective barrier height. The shortest transition wavelength, less than 1.1 μm, was reported in a GaN/AlN MQW [5]. ZnSe/BeTe MQW can be grown on a GaAs substrate. The conduction band (Γ) discontinuity of ZnSe/BeTe heterointerface is as large as 2.3 eV. However, electron transfer from the upper sub-band (Γ) in ZnSe wells to the X state in indirect-gap BeTe barriers degrades the fast response, because of a long carrier lifetime of the X state. In order to locate the upper sub-band level in the well below the X state of the barrier, a CdS layer is inserted in each well [9]. ZnSe layers are indispensable for growing high-quality BeTe and CdS. The conduction band discontinuity of InGaAs and AlAsSb is about 1.6–1.75 eV [12]. Insertion of a thin AlAs intermediate layer between InGaAs and AlAsSb is useful for keeping crystal quality high. The coupled quantum well structure is effective in shortening the absorption recovery time, as well as shortening the transition wavelength [11, 14].

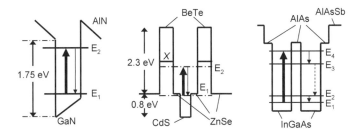

Figure 5.2 Conduction band structures of quantum wells (QW) in which the intersub-band absorption at $\lambda \sim 1.55\,\mu m$ is achieved. (a) GaN/AlN QW, (b) (CdS/ZnSe)/BeTe QW, and (c) InGaAs/AlAs/AlAsSb coupled quantum well (CQW)

5.1.2 Matrix Element

The matrix element of the interband and intersub-band transitions can be approximated by

$$\langle u_{\alpha'}\phi_{j,\alpha'} \,|H_{\mathrm{I}}| \,u_{\alpha}\phi_{i,\alpha}\rangle = \begin{cases} \langle u_{\alpha'} \,|H_{\mathrm{I}}| \,u_{\alpha}\rangle_{cell} \langle \phi_{j,\alpha'} \,|\phi_{i,\alpha}\rangle & (\alpha' \neq \alpha) \\ \langle \phi_{j,\alpha} \,|H_{\mathrm{I}}| \,\phi_{i,\alpha}\rangle & (\alpha' = \alpha) \end{cases} \qquad (5.2)$$

where u_{α} is the periodic part of the Bloch function for the band α, $\phi_{j,\alpha}(z)$ is the envelope function of the jth sub-band of the band α, z is the coordinate normal to the well, and H_{I} is the interaction Hamiltonian [17]. Here, we will not go into the details of the matrix element for the interband transition ($\alpha' \neq \alpha$), which can be obtained by the $\boldsymbol{k} \cdot \boldsymbol{p}$ perturbation theory [18]. In the case of ISBT in the conduction band ($\alpha' = \alpha = c$), the transition occurs between two different envelope states $\phi_{j,c}(z)$. Hereafter, we omit the subscript c. Because of the symmetry of $\phi_j(z)$, the ISBT is allowed only for the light with the electric field normal to the well (E_z). The interaction Hamiltonian can be expressed as $H_{\mathrm{I}} = \boldsymbol{\mu} \cdot \boldsymbol{E} = \mu_z E_z$, where $\mu_z = ez$ is the dipole moment, and e is the electron charge. Therefore, the dipole moment of the transition between ith and jth sub-bands is represented by [1]:

$$\mu_{ji} = \langle \phi_j \,|\mu_z| \,\phi_i\rangle = e \int \phi_j(z)^* z \phi_i(z) dz \qquad (5.3)$$

Since $\mu_z = ez$ is an odd function of z, the transition between the sub-bands with the same parity is forbidden in a symmetric well. This parity selection rule is opposite to that of the interband transition. The dipole moment μ_{ji} increases with the well width (in proportion to L_{w} if the barrier height were infinite) almost independent of materials. The dipole moment for the transition between the adjacent sub-bands is much larger than that for the other transitions.

5.1.3 Saturable Absorption

The absorption coefficient from the lower sub-band i to the upper sub-band j in a homogeneous QW is expressed as [19]:

$$\alpha(\hbar\omega) = \frac{\mu_{ji}^2 (N_i - N_j)}{c_0 \varepsilon_0 n} \frac{\hbar\omega_{ji}/\tau_{\mathrm{ph}}}{\left(\hbar\omega_{ji} - \hbar\omega\right)^2 + \left(\hbar/\tau_{\mathrm{ph}}\right)^2} \qquad (5.4)$$

where N_i is the carrier density in the ith sub-band, c_0 is the velocity of light in vacuum, ε_0 is the electric permittivity in vacuum, n is the refractive index, $\hbar\omega_{ji}$ is the energy separation between the two sub-bands, and τ_{ph} is the dephasing time.

Here, we assume that the energy separation between the two lowest sub-bands is large enough for all the carriers to locate in the lowest sub-band. Then, the unsaturated absorption coefficient at the resonant wavelength is given by:

$$\alpha_0 = \frac{\mu_{21}^2 \omega_{21} N_t \tau_{ph}}{\hbar c_0 \varepsilon_0 n} \tag{5.5}$$

where N_t is the total carrier density in the well. When some carriers are excited to the upper sub-band by light with an intensity of I, the absorption coefficient is reduced. In equilibrium, N_2 satisfies:

$$\frac{dN_2}{dt} = \frac{\alpha I}{\hbar\omega_{21}} - \frac{N_2}{\tau_{21}} = 0 \tag{5.6}$$

where τ_{21} is the intersub-band carrier relaxation time. The saturation intensity I_s is defined as the intensity at which $\alpha(I)$ is equal to $\alpha_0/2$. At I_s, $N_1 = 3N_t/4$ and $N_2 = N_t - N_1 = N_t/4$. From Equation (5.6), I_s is derived as:

$$I_s = \frac{\hbar^2 c_0 \varepsilon_0 n}{2\mu_{21}^2 \tau_{21} \tau_{ph}} \tag{5.7}$$

As discussed in Chapter 1, this relationship can also be derived form the third-order nonlinear susceptibility. This equation shows a trade-off between the response time and the switching energy.

5.1.4 Absorption Recovery Time

The relaxation process of the excited electron is schematically shown in Figure 5.3 [20]. At room temperature, longitudinal optical (LO) phonon scattering is the dominant energy relaxation process [3]. In the case of near-infrared ISBT, excited electrons are scattered to the high-energy state of the lower sub-band. Because of the band nonparabolicity, the energy separation of the two sub-bands is smaller for a larger wavenumber $k_{//}$. Therefore, the electrons in the high-energy state of the lower sub-band cannot absorb the signal light. The electrons lose the energy by emitting several LO phonons. The carrier cooling time is generally comparable to or longer than τ_{21}. When the carrier density is high, however, the electron distribution near the sub-band edge of the lower sub-band recovers through electron-electron scattering (thermalization). Generally speaking, the electron–electron scattering time τ_{ee} is about one order of magnitude less than τ_{21}. Therefore, the absorption recovery time at a shorter wavelength (representing the transition for $k_{//} \sim 0$) is dominated by τ_{21}. On the contrary, the carrier distribution near the Fermi level does not recover until the carrier temperature recovers. Therefore, an additional slower time constant component appears in the absorption recovery at a longer wavelength [21, 22]. When the barrier material is an indirect semiconductor, a barrier sub-band in L or X valley may be formed below the upper sub-band level of the wells. In this case, some of the excited electrons are trapped by the barrier sub-band with a long lifetime, which seriously slows down the absorption recovery [9]. $Al_x Ga_{1-x}N$ has a direct bandgap for entire composition ($x = 0 - 1$). In the case of ZnSe/BeTe, the upper sub-band level in the well should be lowered by the insertion of CdS. In the case of InGaAs-based QWs, formation of the indirect-gap barrier sub-band is avoided by adopting direct-bandgap AlAsSb as the barrier material.

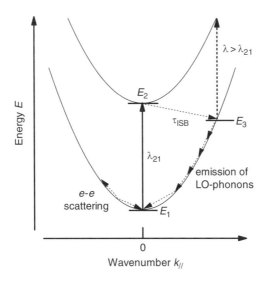

Figure 5.3 The intersub-band relaxation process and a three-level model. E_1 and E_2 represent the sub-band edge energy levels of the lower and the upper sub-bands, respectively. E_3 represents the higher energy states of the lower sub-band, which cannot contribute the transition because of the band nonparabolicity [20]

For an infinite well, the intersub-band LO-phonon scattering rate can be approximated by [3]:

$$W_{21} = \frac{1}{2} W_0 \left(\frac{\hbar\omega_{LO}}{E_1} \right)^{1/2} \left[\frac{1}{4 - (\hbar\omega_{LO}/E_1)} + \frac{1}{12 - (\hbar\omega_{LO}/E_1)} \right] \qquad (5.8)$$

where

$$W_0 = \frac{e^2}{4\pi\hbar} \left(\frac{2m^*\omega_{LO}}{\hbar} \right)^{1/2} \left[\frac{1}{\varepsilon_\infty} - \frac{1}{\varepsilon_s} \right] \qquad (5.9)$$

is the basic rate, ω_{LO} is the angular frequency of the LO-phonon; and ε_∞ and ε_s are the optical and the static dielectric permittivities, respectively. The material constants of GaN, CdS, and InGaAs are compared in Table 5.1. The intersub-band relaxation times estimated from $\tau_{ISB} \sim 1/W_{21}$ are shown in Figure 5.4 as functions of the transition wavelength [23–25]. Some experimental data for the absorption recovery times (τ_{rec}) are also shown in the Figure. At $\lambda \sim 1.55\,\mu m$, the absorption recovery times were reported to be 110–400 fs for GaN/Al(Ga)N MQWs [6, 22, 26–29], 160–270 fs for II–VI MQWs [9, 21], and 690 fs in InGaAs/AlAsSb CQWs [12], respectively. Equation (5.8) overestimates the scattering rate for the near-infrared ISBT, because the barrier height is assumed to be infinite, and interface (IF) phonon modes [30], whose contribution is important in a narrow QW, are ignored. The thick line in Figure 5.4 shows τ_{ISB} in an $Al_{0.8}Ga_{0.2}N/GaN$ QW calculated by a numerical calculation model [24, 31], which fits the experimental results better ($\tau \sim 110$ fs when $\lambda = 1.55\,\mu m$). The response of InGaAs is an order of magnitude slower than those of GaN and CdS because of weaker interaction between electrons and LO-phonons. However, the apparent absorption recovery time for the 1–4 transition can be shortened in coupled double QWs [12], because the relaxation times τ_{43} and τ_{21} are much faster than τ_{41}. The repetition rate is limited by τ_{42}, τ_{41}, τ_{32}, and $\tau_{31}(\sim \text{ps})$.

Table 5.1 Comparison of material constants related to the carrier relaxation processes in GaN, CdS, and InGaAs [23–25]

Material	m^*/m_0	$\varepsilon_s/\varepsilon_0$	$\varepsilon_\infty/\varepsilon_0$	$\hbar\omega_{LO}$	W_0
GaN	0.20	9.5	5.35	88 meV	$1.2 \times 10^{14}\,\text{s}^{-1}$
CdS	0.19	10.3	5.2	38 meV	$9.1 \times 10^{13}\,\text{s}^{-1}$
InGaAs	0.042	14.1	11.6	36 meV	$6.7 \times 10^{12}\,\text{s}^{-1}$

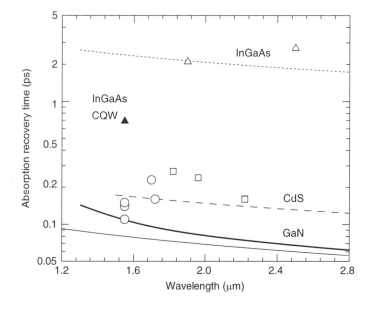

Figure 5.4 Comparison of the measured absorption recovery time [6, 9, 12, 21, 22, 26–30] and the calculated intersub-band relaxation time for GaN (solid lines and circles) [23], CdS (dashed line and squares), and InGaAs (dotted line and triangles) [17]. The thin lines were the estimations calculated using Equation (5.8). The thick solid line was obtained by a more rigorous numerical calculation [24, 25]

5.1.5 Dephasing Time and Spectral Linewidth

Dephasing time, τ_{ph}, is the mean time in which the electrons lose coherence through various scattering processes. Equation (5.7) suggests that a longer τ_{ph} is preferable for smaller saturation intensity. On the other hand, it relates to the homogeneous spectral linewidth as:

$$\Delta\nu_{homo} = 1/(\pi\tau_{ph}) \tag{5.10}$$

Ultrafast response is achievable only within the homogeneous linewidth. A narrower $\Delta\nu_{\text{homo}}$ limits the response for a short pulse with broad power spectrum. In order to allocate different wavelengths to the signal and the control pulses, a wider $\Delta\nu_{\text{homo}}$ is required. For example, a homogeneous bandwidth of about 100 meV is required to cope with 100-fs pulses.

The dominant intraband carrier scattering processes are electron–electron scattering, ionized impurity scattering, interface scattering, and LO-phonon scattering [32]. If we ignore interface scattering, the dephasing time can be approximated by:

$$\tau_{\text{ph}} = \left[\tau_{\text{ee}}^{-1} + \tau_{\text{ii}}^{-1} + \tau_{\text{ep}}^{-1} + (2\tau_{\text{ISB}})^{-1}\right]^{-1} \tag{5.11}$$

where τ_{ee}, τ_{ii}, and τ_{ep} are the electron–electron scattering time, the ionized impurity scattering time, and the intrasub-band LO-phonon scattering time, respectively.

The electron–electron scattering time τ_{ee}, the ionized impurity scattering time τ_{ii}, and the intrasub-band LO-phonon scattering time τ_{ep} were calculated for an $Al_{0.8}Ga_{0.2}N/GaN$ (6 MLs) QW [24, 25]. Carriers were assumed to be in thermal equilibrium at $T = 300$ K. The results are shown in Figure 5.5 as functions of the carrier density N. The electron–electron scattering time τ_{ee} is about 10 fs when $N = 10^{17}$–10^{18} cm^{-3}. For lower N, τ_{ee} increases because the probability of electron collision is reduced. For higher N, τ_{ee} increases because of the increased probability that the final state is already occupied (exclusion principle). The screening also contributes to the increase in τ_{ee} for higher N. The ionized impurity scattering time shows a similar tendency. The LO-phonon emission rate and the LO-phonon absorption rate of an electron in GaN are about 1×10^{14} s^{-1} and 3×10^{12} s^{-1}, respectively, if the final state is available. In

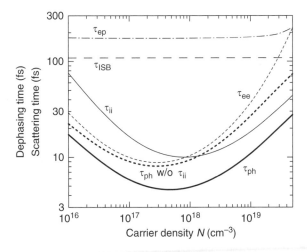

Figure 5.5 Calculated electron–electron scattering time τ_{ee} (thin dotted line), ionized impurity scattering time τ_{ii} (thin solid line), intrasub-band LO-phonon scattering time in the lowest sub-band τ_{ep} (thin dot dashed line), intersub-band relaxation time τ_{ISB} (thin dashed line), and dephasing time τ_{ph} (thick solid line) in GaN/AlN MQW as functions of carrier density [24, 25, 35]. The thick dotted line denotes the dephasing time when the ionized impurity scattering can be ignored. Electrons were assumed to be in thermal equilibrium at $T = 300$ K [25]. (Reproduced by permission of © 1998 International Society for Optical Engineering (SPIE))

thermal equilibrium, however, since the kinetic energy $E_{//}$ of most of the electrons is lower than the LO-phonon energy $\hbar\omega_{LO}$, they cannot emit an LO-phonon. Therefore, the averaged intrasub-band LO-phonon scattering time is rather long ($\sim 200\,\text{fs}$). The dephasing time of GaN with a carrier density higher than $10^{19}\,\text{cm}^{-3}$ was estimated to be 10–40 fs. Nonequilibrium carrier distribution under strong excitation may moderate the effect of the exclusion principle, which may shorten the dephasing time for higher N. Another scattering mechanism, such as interface roughness scattering, may further shorten the dephasing time. In the case of GaN ($\tau_{\text{ph}} = 10\text{–}40\,\text{fs}$), the homogeneous linewidth is 30–130 meV. The short dephasing time assures broadband (>100 nm) operation of GaN-based ISBT gate switches, although it leads to rather high switching pulse energy. Contribution of the ionized impurity scattering can be reduced by means of modulation doping in the barriers. When ionized impurity scattering can be ignored, dephasing time is increased. The electron–electron scattering rate is proportional to m^*/ε_∞^2 [32]. The dephasing time of InGaAs is an order of magnitude longer than that of GaN.

5.1.6 Gate Operation in Waveguide Structure

Since ISBT is allowed only for light polarized normal to the wells, gate switches are to be realized in a low-loss waveguide for TM-mode. Assuming that the background loss and the coupling loss are almost independent of polarization, the difference in transmittances for TM- and TE-modes is considered to be the absorption due to ISBT. To achieve a higher extinction ratio, this polarization dependent loss (PDL) should be higher. In actual waveguides, however, experimentally observable PDL is limited to a certain level (typically $\sim 30\,\text{dB}$) due to mode conversion and misalignment. When the unsaturated loss due to ISBT is greater than this background level, saturation is hardly observed. Therefore, the MQW and the waveguide are designed so that the unsaturated PDL becomes 20–25 dB.

The response of GaN-based ISBT gates for ultra-short optical pulses has been analyzed using a one-dimensional, finite-difference, time-domain (FDTD) model combined with the rate equations describing the intersub-band carrier dynamics [20, 33]. Figure 5.6 shows the assumed waveguide structures: a ridge waveguide (RWG) structure and a high-mesa (HM)

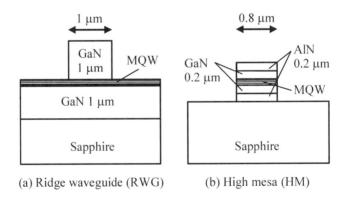

(a) Ridge waveguide (RWG)   (b) High mesa (HM)

Figure 5.6 Waveguide structure assumed in the FDTD simulation of GaN/AlN ISBT gate switches. (a) Ridge waveguide (RWG), and (b) high-mesa (HM) waveguide [33]. (Reproduced by permission of © 2005 IEICE)

structure. Figure 5.7 shows the absorption saturation characteristics for a 250-fs Gaussian pulse resonant to ISBT. The saturation pulse energy is given by:

$$E_s = I_s A_{eff} t_p \tag{5.12}$$

where A_{eff} and t_p are the effective mode area and pulse width, respectively. Saturation pulse energy in an HM waveguide is smaller than that in a RWG because of a smaller A_{eff}. The saturation intensity in waveguides is higher than the value given by Equation (5.7), because of the power attenuation in the waveguide [33]. Figure 5.8 shows the pulse width dependence of the saturation pulse energy and the saturation pulse intensity. The saturation pulse energy is minimized when the homogeneous absorption spectrum and the power spectrum of the pulse match each other. The optimum pulse width is around 100 fs.

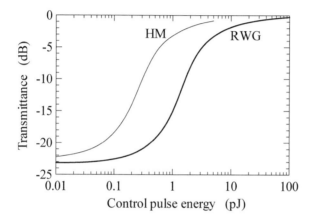

Figure 5.7 Absorption saturation characteristics of the ridge waveguide (RWG) and the high-mesa (HM) waveguide shown in Figure 5.6 [33]. (Reproduced by permission of © 2005 IEICE)

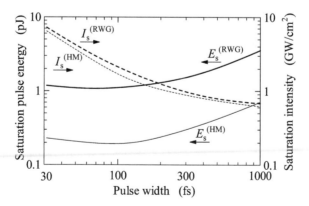

Figure 5.8 Pulse width dependence of the saturation pulse energy (solid lines) and the effective saturation intensity (dotted lines) for the ridge waveguide (RWG) and the high-mesa (HM) waveguide [33]. (Reproduced by permission of © 2005 IEICE)

The saturation pulse energy in actual waveguides is rather high compared with the prediction in Figure 5.8 because of inhomogeneous broadening, small overlap of the mode field and the MQW, and background loss. In GaN/AlN waveguides, lots of dislocations run normal to the wells. The dislocations trap the charges and serve as a polarization filter [34], causing serious excess PDL. Reduction of the excess PDL is crucial in observing the saturation. In GaAs-based materials, two-photon absorption (TPA), which occurs in the same intensity range as I_s, destructively interferes with the saturation of ISBT [13]. Suppression of TPA is important for reducing the switching energy. For (CdS/ZnSe)/BeTe MQWs, ZnMgBeSe can be utilized as a cladding lattice-matched to the GaAs substrate [21].

Figure 5.9 shows the calculated response for a 10-pJ input pulse at the input edge of the RWG [33]. N_1 and N_2 denote the densities of carriers contributing to the transition in the lower and the upper sub-bands, respectively. N_3 represents the density of carriers in higher energy states of the lower sub-band, which cannot contribute to absorption of signal light because of the band nonparabolicity (See Figure 5.3). The intersub-band relaxation time τ_{23} and the intrasub-band carrier cooling time τ_{31} were assumed to be 150 fs and 200 fs, respectively. The absorption is proportional to $(N_1 - N_2)$, and recovers within 1.5 ps after the pulse peak passes.

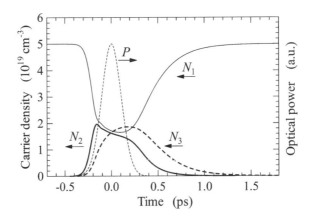

Figure 5.9 Changes in electron densities at the input edge of the ridge waveguide (RWG). The wavelength, the width, and the energy of the pulse were assumed to be 1.57 μm, 250 fs, and 10 pJ, respectively [33]. (Reproduced by permission of © 2005 IEICE)

5.2 GaN/AlN ISBT Gate

The superior properties of GaN-based ISBT gates are as follows: (a) ultrafast absorption recovery ($\tau \sim 100$ fs), which enables pattern-effect free 1-Tb/s operation; (b) short wavelength operation up to ~ 1.1 μm; (c) broadband operation (~ 200 nm), which allows the switching of 100-fs pulses, (d), free from TPA; (e) potential for high-temperature (> 100 °C) and high-power operation; and (f) free from toxic materials [35]. GaN/AlN layers are usually grown on *c*-plane sapphire substrate. Lattice-mismatch between GaN and AlN is about 2.5 %. Nitride semiconductors have large piezoelectric effects and spontaneous polarization [36]. Built-in fields in polar GaN/Al(Ga)N MQWs are of the order of MV/cm [16, 37], which have drastic effects on the intersub-band absorption wavelength. The transition wavelength depends on not only the well width, but also the barrier thickness, the structure of underlying layers, and the substrate. The biggest challenge for GaN-based ISBT gates is the switching energy. As Equation (5.7) suggests, the switching energy tends to be high in ultrafast switches. In the case of GaN, polarization-dependent loss caused by edge dislocations further increases the switching energy. Reduction of dislocation density is crucial for realizing a practical ultrafast all-optical gate switch.

5.2.1 Absorption Spectra

The calculated transition wavelengths of GaN/Al(Ga)N MQWs are compared with some experimental data in Figure 5.10 [16, 37]. In this calculation, the built-in field was assumed to be uniform, and the sub-band levels were calculated by the transfer matrix technique [38]. The conduction band discontinuity of GaN and AlN was assumed to be 1.76 eV, considering a deformation potential (-9 eV for AlN [39]). The band nonparabolicity parameter was assumed to be 0.187 eV^{-1} [40]. The thick and thin lines denote GaN/AlN MQWs and GaN/Al$_{0.65}$Ga$_{0.35}$N MQWs, respectively. The solid circles and the open triangles show the experimental data for AlN/GaN MQWs grown by molecular beam epitaxy (MBE) [6] and for GaN/AlGaN MQWs grown by metal-organic chemical deposition (MOCVD) [16], respectively. Near-infrared ISBT of $\lambda < 2\,\mu$m has not been achieved in MOCVD-grown samples, because of the inferior heterointerface and lower carrier concentration ($<10^{18}$ cm^{-3}) [16, 41–43]. From

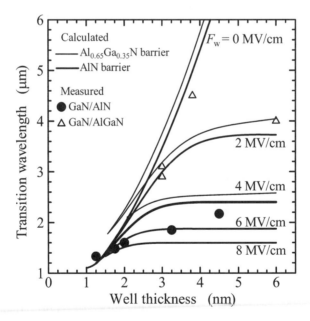

Figure 5.10 Effect of built-in field on intersub-band transition wavelength [16, 37]. The thick and thin lines show the transition wavelength for GaN/AlN MQWs and GaN/Al$_{0.65}$Ga$_{0.35}$N MQWs, respectively, which were calculated assuming a uniform built-in field in GaN well. The solid circles and the open triangles represent the measured data for AlN/GaN MQWs grown by molecular beam epitaxy (MBE) [6] and for GaN/AlGaN MQWs grown by metal-organic chemical deposition (MOCVD) [16], respectively [37]. (Reproduced by permission of © 2003 The Institute of Pure and Applied Physics (IPAP))

the figure, the effective field strengths in the well were estimated to be 5–10 MV/cm and 0–3 MV/cm for GaN/AlN MQWs and GaN/Al$_{0.65}$Ga$_{0.35}$N MQWs, respectively.

The dotted lines in Figure 5.11 show the absorption spectra of MBE-grown GaN/AlN MQWs [6]. The thicknesses of GaN wells were 4.5 nm (Sample A), 3.3 nm (B), 2.0 nm (C), 1.75 nm (D), and 1.2 nm (E), respectively. The thickness of AlN barrier was 4.6 nm, and the numbers of wells (N_w) were 30–120. The absorption peak wavelengths were 2.17 μm (A), 1.85 μm (B), 1.60 μm (C), 1.48 μm (D), and 1.33 μm (E), respectively. The sub-bands and the potentials of these samples were calculated self-consistently under periodic boundary condition [37], which is a good approximation when $N_w \geq 30$. The solid lines in Figure 5.11 show the absorption spectra calculated considering the experimental arrangement. To get better fits to the measured transition wavelengths, two-monolayer composition grading regions were assumed at the heterointerfaces, and half of the carriers were assumed to be trapped by dislocations. For Samples C, D, and E, a well thickness fluctuation of ±1 ML was assumed in order to simulate the shoulders observed in the measured spectra. The ratios of the +1-ML and −1-ML portions were both assumed to be 25 %. A good fit was obtained.

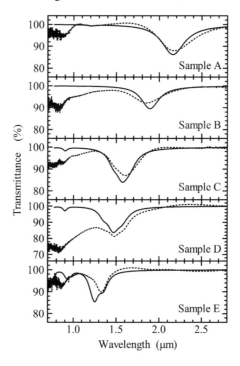

Figure 5.11 Calculated (solid lines) and measured (dotted lines) transmittance spectra of the GaN/AlN MQWs. The thicknesses of GaN wells were 4.5 nm (Sample A), 3.3 nm (B), 2.0 nm (C), 1.75 nm (D), and 1.2 nm (E), respectively. The increase in loss in the shorter wavelength for some measured data is caused by surface scattering [37]. (Reproduced by permission of © 2003 The Institute of Pure and Applied Physics (IPAP))

Figure 5.12(a), (b), and (c) show the calculated sub-band structures of Samples A, C, and E, respectively. In Sample A, the effective well width for the lower sub-bands is determined by the tilt of the potential, independent of the well thickness. The built-in field drastically shortens the transition wavelength in thick QWs. In Sample C, the second sub-band is confined by the heterointerfaces on both sides. In this case, the transition wavelength mainly depends on the well width, rather than the built-in field. The second sub-band in Sample E is confined by the tilted potential of the barrier on the left side. Thus, effective barrier height is raised by the built-in field in polar nitride QWs. An absorption peak wavelength shorter than 1.6 μm has not been reported in nonpolar GaN/AlN MQWs, where the barrier height is not enhanced [44, 45].

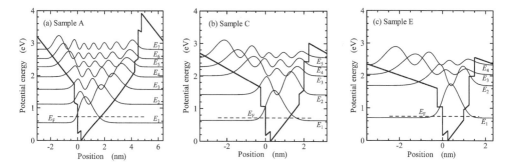

Figure 5.12 Calculated conduction band structures for (a) Sample A, (b) Sample C, and (c) Sample E. The squared envelope functions are shown. Two-monolayer graded regions were assumed in the heterointerfaces. The left side of the figure is the surfaced side [37]. (Reproduced by permission of © 2003 The Institute of Pure and Applied Physics (IPAP))

Since the potential is asymmetric, the 1–3 transition is allowed in polar nitride MQWs. Clear 1–3 transitions at $\lambda = 1.07$–2.0 μm were observed by some groups [41, 46, 47]. Shorter wavelength transitions were predicted in GaN-based CQWs (see Figure 5.13) [37]. Absorption spectra peaking at 1.2–2.3 μm were reported in GaN/AlN CQWs [48].

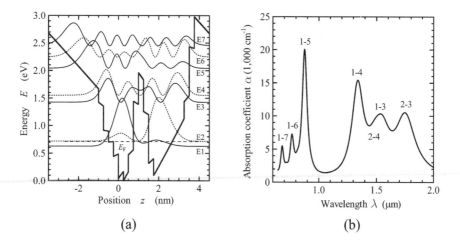

Figure 5.13 A design of GaN/AlN coupled quantum wells to enhance shorter wavelength intersub-band transition. (a) Conduction band structure, and (b) calculated absorption [37]. (Reproduced by permission of © 2003 The Institute of Pure and Applied Physics (IPAP))

5.2.2 Saturation of Absorption in Waveguides

In nitride layers grown on a sapphire substrate, a lot of edge dislocations run perpendicularly to the substrate. The dislocation density in MBE-grown samples is typically of the order of $10^{10}\,cm^{-2}$, which is one order of magnitude higher than that in MOCVD-grown samples. Electric charges trapped by the dislocations cause excess polarization-dependent loss (PDL) for the TM mode [34]. Actually, PDL of 8–15 dB/mm was observed in GaN waveguides without MQWs. The excess PDL depends on the dislocation density and the carrier density, but was almost independent of the optical pulse energy. Reduction in edge dislocations is indispensable for reducing the saturation intensity in nitride waveguides.

Three types of ridge waveguides with GaN/AlN ISBT layers were fabricated [28, 29, 49]. Figure 5.14 shows the cross-sections. Samples A and B were solely grown by MBE. Sample B has a multiple intermediate layer (MIL) consisting of 10 pairs of unintentionally-doped GaN (40 nm) and AlN (10 nm) [28], which serve as a dislocation filter [50]. Sample C is grown on a 0.8-μm thick MOCVD-grown GaN template [29]. The MQW of Sample B consisted of 10 pairs of Si-doped ($5 \times 10^{19}\,cm^{-3}$) GaN (2 nm) and AlN (3 nm). The absorption peak wavelength

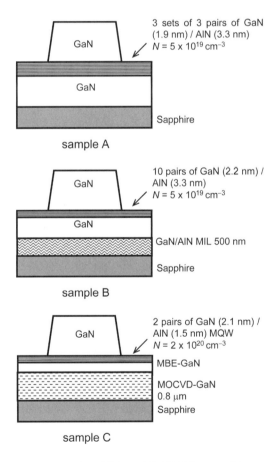

Figure 5.14   Cross-sectional views of the fabricated GaN/AlN waveguide structures [49]. (Reproduced by permission of © 2006 IEEE)

and the linewidth were 1.75 μm and 120 meV, respectively. In Sample C, the number of wells N_w was reduced to two and the barrier thickness was reduced to 1.5 nm to suppress generation of edge dislocations in the MQW layer. To compensate for the reduction in N_w, the Si-doping level was increased to 2×10^{20} cm^{-3}. The total thickness of the nitride layers, the ridge height, ridge width, and waveguide length were about 2 μm, 1 μm, 1–2 μm, and 400 μm, respectively. The chips were AR-coated and housed in a module with fiber pigtails as shown in Figure 5.15.

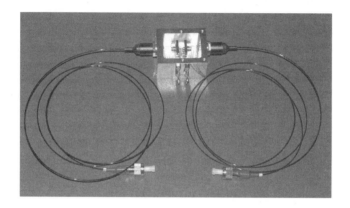

Figure 5.15   Photograph of a fabricated GaN ISBT gate module with fiber pigtails

The absorption saturation characteristics at $\lambda = 1.55$ μm were measured by utilizing optical pulses generated by an optical parametric oscillator (OPO) excited by a mode-locked Ti:sapphire laser with a repetition rate of 80 MHz. The pulse width was 130 fs. TE-mode insertion loss including the coupling loss was 8.6, 7.8, and 6.6 dB for Samples A, B, and C, respectively. Figure 5.16 shows the measured polarization-dependent loss (PDL). At a fiber input pulse energy of 100 pJ, PDL was decreased by 3 dB, 5 dB, and 10 dB for Samples A, B, and C, respectively. The 3-dB saturation pulse energies were 100 pJ, 54 pJ, 25 pJ, respectively. In the waveguides without an MQW layer similar to Samples A, B and C, excess PDL was measured to be 18, 10, and 1 dB/mm, respectively. The excess PDL in the samples with a Si-doped MQW is considered to be higher. For example, the excess PDL of Sample C was estimated to be 4.5–5 dB from the transmittance at the wavelength outside the absorption spectrum.

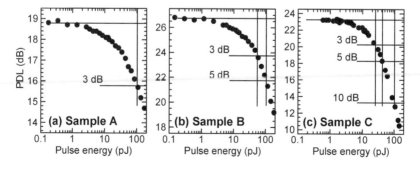

Figure 5.16   Absorption saturation characteristics of Samples A, B, and C [49]. (Reproduced by permission of © 2006 IEEE)

In AlN-based waveguides, the same amount of absorption can be achieved with a lower Si-doping concentration compared with GaN-based waveguides, because of a stronger carrier confinement. It is advantageous for reducing the excess PDL caused by the carriers trapped by dislocations. The 10-dB saturation pulse energy of 100 pJ was achieved in an AlN high-mesa waveguide [51], in spite of inferior etched side wall. The shortest ISB absorption wavelength ($\sim 1.35 \, \mu m$) in waveguides was also realized in an AlN-based waveguide, because the built-in field in GaN wells is enhanced compared with that in GaN-based waveguides [52]. In a rib waveguide structure with an AlN lower cladding and a GaN mesa, an excess PDL of 2 dB/mm and a 3-dB saturation pulse energy of less than 10 pJ were reported [53].

As shown in Figure 5.17, the 10-dB switching pulse energy is predicted to be 10 pJ and 1 pJ in an improved ridge waveguide and in an ideal nitride high-mesa waveguide, respectively [33]. Reductions in the coupling loss and the scattering loss are crucial as well as the reduction in the excess PDL due to the dislocations. The switching pulse energy may be further reduced by adopting a slot waveguide [54]. Even if the switching energy is reduced to 1 pJ, the power consumption amounts to 1 W at 1 THz. Nitride semiconductors can withstand such power consumption as long as they are formed on a material with a good thermal conductivity. When 1-Tb/s OTDM technology matures in future, the function of a photonic node will be integrated onto a photonic integrated circuit. Polarization dependence of ISBT will not be a fatal problem on such a photonic circuit integrated with a pulse laser.

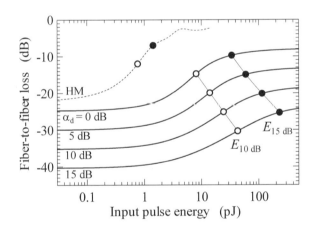

Figure 5.17 Calculated saturation characteristics of improved GaN-based ISBT gates. The thick lines represent the saturation characteristics of ridge waveguide gates with the excess PDL of $\alpha_d = 0, 5, 10,$ and 15 dB. The thin line represents the characteristics of a high-mesa gate with $\alpha_d = 0$ dB. The open and solid circles denote the 10- and 15-dB switching pulse energy [33]. (Reproduced by permission of © 2005 IEICE)

5.2.3 Ultrafast Optical Gate

Optical gate operations of Samples B and C were investigated by utilizing the pump and probe technique [28, 29, 49]. In Sample B, the 1.55-μm signal pulses were modulated by 1.7-μm, 120-pJ control pulses as shown in Figure 5.18(a) [28]. The FWHM of the response was 360 fs, which was limited by the convolution of the two pulses. The control pulse width was broadened to 230 fs in the 1-m fiber pigtail due to the dispersion and nonlinearity. A rather small modulation ratio (~2.4 dB) is due to the excess PDL. In Sample C, the 1.55-μm signal pulses were switched by 11.5 dB by 1.7-μm, 150 pJ control pulses as shown in Figure 5.18(b)

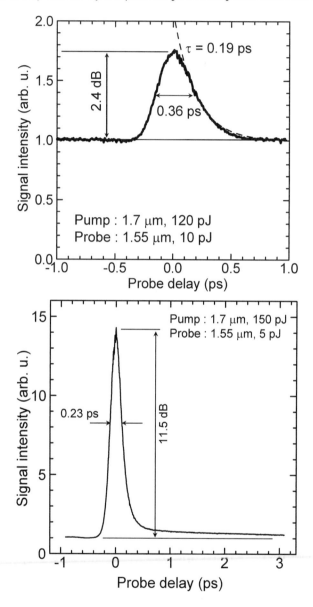

Figure 5.18 Pump and probe responses of GaN/AlN ISBT gate switches. (a) Response of Sample B [28]. The pulse energies of the 1.7-μm pump and 1.55-μm probe pulses were 120 pJ and 10 pJ, respectively. (b) Response of Sample C [29]. The pulse energies of the pump and probe pulses were 150 pJ and 5 pJ, respectively [49]. (Reproduced by permission of © 2006 IEEE)

[29]. The gate width was reduced to 230 fs, since the broadening of the control pulse was suppressed by shortened fiber pigtails (50 cm).

As shown in Figure 5.19, pattern-effect-free optical modulation corresponding to 1.5 Tb/s was achieved in Sample B [49, 55]. In the case of Sample C, however, a slow recovery component ($\tau \sim 2$ ps) was observed in addition to the fast recovery ($\tau_{ISB} \sim 110$ fs). Pattern effects were observed above 500 Gb/s as shown in Figure 5.20 [56]. In Sample C, the barrier thickness was reduced to suppress the generation of dislocation in the MQW. The slow recovery is attributable to the tunneling of excited electrons to the adjacent bulk GaN layer.

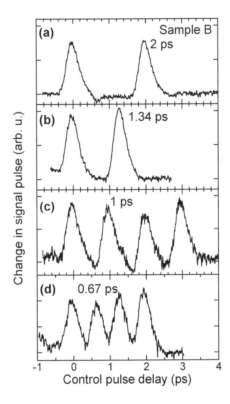

Figure 5.19 Changes in the transmitted signal intensity as a function of signal delay for Sample B [55]. The signal pulse intervals were (a) 2 ps, (b) 1.34 ps, (c) 1 ps, and (d) 0.67 ps. The pump pulse energy was 100 pJ, and the signal pulse energy was 4 pJ [49]. (Reproduced by permission of © 2006 IEEE)

5.3 (CdS/ZnSe)/BeTe ISBT Gate

The reports of ISBT in ZnSe/BeTe superlattices (SLs) with a conduction-band offset (CBO) of 2.3 eV have created opportunities for applying II–VI wide gap semiconductors to devices in the near-infrared spectral region [57]. However, a slower relaxation process with a time constant of a few ps merges and becomes a dominant decay process in ISBT with wavelength $\lambda < 2\,\mu m$ due to Γ(ZnSe)–X(BeTe) electron transfer [58]. This limits its application as an ultrafast all-optical device in the fiber-optic communication wavelength region. In order to circumvent

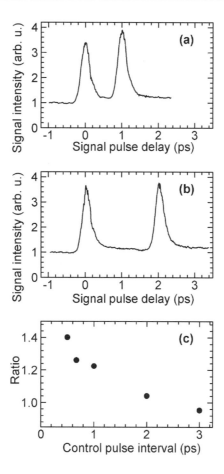

Figure 5.20 Temporal change in transmitted signal intensity with dual pump pulses with intervals of (a) 1 ps and (b) 2 pJ for Sample C [56]. (c) Ratio of the second transmitted signal pulse to the first pulse as a function of the control pulse interval [49]. (Reproduced by permission of © 2006 IEEE)

this disadvantage in ZnSe/BeTe SLs, a CdS-ZnSe-BeTe material system has been developed to suppress the slow relaxation of carriers [9]. A huge CBO of up to 3.1 eV in this material system enables the tuning of ISBT over a wide range to realize absorption at a wavelength of around 1.55 μm. The advantageous features such as a large band gap excluding the effect TPA and synthesizable large refractive index contrast for strong optical field confinement in the waveguide, enable us to realize a low-switching-energy optical gate despite its very fast response speed.

5.3.1 Growth of CdS/ZnSe/BeTe QWs and ISBT Absorption Spectra

CdS/ZnSe/BeTe QWs were grown on a GaAs substrate using the molecular beam epitaxy method. The well layer (CdS/ZnSe) is doped homogeneously with $ZnCl_2$ for electron population at the first sub-band. Each sub-layer has an obvious difference in optimal growth temperature (T_s), such as CdS ~ 150–200, ZnSe ~ 260–280, and BeTe ~ 320–$400\,°C$, So, there is a trade-off for the optimum growth T_s of (CdS/ZnSe)/BeTe QWs. Figure 5.21(a) shows (002) $\omega/2\theta$ scans recorded from a series of (CdS/ZnSe)BeTe QWs grown at different T_s [10]. A perfect structure with a narrow linewidth (34 arcsec) has been obtained at

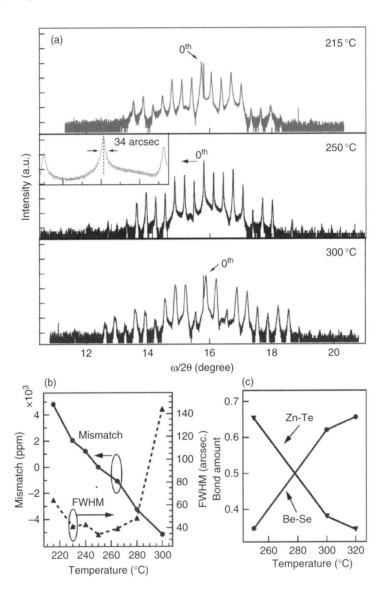

Figure 5.21 (a) (002) $\omega/2\theta$ scans of (CdS/ZnSe)/BeTe QWs grown at different T_s. The inset at upper left shows the a_{QW} matching the GaAs substrate (dotted line). A clear interference fringe is observed in this curve; (b) the dependence of lattice mismatch to the GaAs substrate and FWHMs in XRD curves on the Ts; (c) the amount of chemical bonds in the transition region in ZnSe/BeTe heterostructure vs Ts, obtained by XRD simulation. Be–Se bonds are preferentially formed at high T_s [10]. (Reproduced by permission of © 2006 American Institute of Physics (AIP))

$T_s = 250\,^\circ\text{C}$, while structures degrade at higher or lower T_s. The satellites approach 12 orders in a wide angle range, and the average lattice constant, a_{QW}, matches the GaAs substrate, as indicated in the inset. Clear interference fringes can be observed around the zero-th peak in the samples grown at $T_s \sim 230\text{–}265\,^\circ\text{C}$, meaning sharp interfaces in (CdS/ZnSe)/BeTe QWs. The lattice mismatch shifts from 0.48 % to -0.52 % with increasing T_s, as shown in Figure 5.22(b); correspondingly this results in strain transfer from compressive to tensile with respect to the GaAs substrate. Using XRD dynamic simulation for ZnSe/BeTe QWs, we find that, in addition to the expected Zn–Te bonds in the ZnSe/BeTe transition region, interfacial Be–Se bonds are present in QWs grown at 250 °C, and the number of Be–Se bonds increases for growth at a higher T_s (Figure 5.23(c)). At the CdS/ZnSe interface, Heske *et al.* have demonstrated that more Zn–S bonds are preferentially formed at higher T_s through soft X-ray emission spectroscopy [59]. The Be–Se and Zn–S bonds induce tensile strain in the (CdS/ZnSe)/BeTe heterostructure.

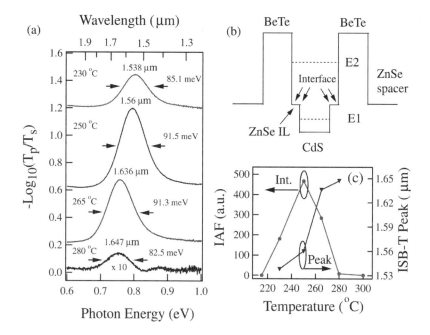

Figure 5.22 (a) Absorption spectra for samples prepared at different T_s, obtained by taking $-$log of the transmission ratio of p-polarized to s-polarized light. The linewidths of absorption spectra are given. (b) Energy band diagram of C B for the (CdS/ZnSe)/BeTe QWs. (c) Dependence of ISBT intensity and wavelength on T_s [10]. (Reproduced by permission of © 2006 American Institute of Physics (AIP))

Shown in Figure 5.22(a) are ISBT spectra for samples grown at different T_s. The dependences of ISBT wavelength and integrated absorption fraction (IAF) in connection with absorption strength on the T_s are summarized in Figure 5.22(c). They show a number of features. First, the wavelength of ISB absorption shifts to the longer side, from 1.538 to 1.647 μm, with increasing T_s. Second, the ISBTs in (CdS/ZnSe)/BeTe QWs only occur in a narrow range of T_s. The 1.56-μm ISB absorption approaches a maximum in this structure grown at 250 °C. It is crucial to grow (CdS/ZnSe)/BeTe QWs by inserting a ZnSe interlayer (IL) [60]. However, it apparently leads to the formation of a complicated interface. As shown in Figure 5.22(b), there are four interfaces due to the introduction of ZnSe IL. Be–Se or Zn–Te bonds in BeTe/ZnSe, as well as Zn–S or Cd–Se bonds in ZnSe/CdS, that exist in the interfacial region. The interfacial configurations have a dominant influence on the strain, structural, optical and electronic properties in a narrow well structure. We explain that variation of ISBT in (CdS/ZnSe)/BeTe QWs results from the enhancement of interface intermixing at higher T_s. The CBO decreases due to the increase of Se–Be bonds at the interfacial region of the ZnSe/BeTe heterostructure. This has already been investigated using photoelectron spectra by Nagelstrasser *et al.* [61]. The offset reduction also occurs in CdS/ZnSe heterostructure at high preparation temperatures [62]. As a result, the CBO in BeTe/ZnSe/CdS will decrease leading to the red shift of ISBT. A high temperature also generates more local states in the transition region due to the enhancement of interdiffusion and atomic exchange. These local states capture electrons and lead to the vanishing ISBT in the (CdS/ZnSe)/BeTe QWs

Figure 5.23(a) shows the ISB absorption of (CdS/ZnSe)/BeTe QWs with nominal well widths of 3.75–14 MLs. The transition wavelength decreases from 3.4 to 1.52 μm with the decrease in the well width. For the first three samples, the wavelengths shift from 3.4 to 2.59 μm with thinning well width. The intensity of the ISB absorption increases six times due to the improvement in structural quality with decreasing CdS thickness from 10 to 5 MLs. In (CdS/ZnSe)/BeTe multi-QWs, lattice relaxation occurs if the CdS ≥ 7 MLs, confirmed by *in situ* RHEED pattern. Then the intensity of ISBT gradually decreased from 2.59 to 1.56 μm due to the decrease in active layer thickness. However, the intensity of 1.52-μm ISB absorption drops drastically with a slight decrease of CdS thickness. The nominal 3.5-MLs (1-ML ZnSe/1.5-ML CdS/1-ML ZnSe) wide wells do not show any ISB absorption. To find out the reason, we carried out photo-induced ISBT (PI-ISBT) measurement on this structure using lock-in techniques combined with a step-scan of FTIR under excitation by a frequency-doubled mode-locked Ti:sapphire laser at $\lambda = 375$ nm. With laser shining, electron-hole pairs are generated in the (ZnSe/CdS/ZnSe) well layer and electrons are injected to ground state (e_1) within QW. We observe the ISBT down to 1.49 μm with a FWHM of 86 meV. It indicates that the e_2 state is still confined in QW and there is no free electron in the well. Referring back to Figure 5.22(b), the total interfacial regions are up to 3 or 4 MLs, which is comparable to the intended well thickness of 3–4 MLs [63]. The interfacial states capture electrons and decrease the electron sheet density, finally resulting in ISB absorption disappearing.

The separation of energy (e_1 to e_2) is numerically calculated using envelop function approximation. Due to the high ISBT energies in (CdS/ZnSe)/BeTe QWs, the nonparabolicity of the bulk conduction band dispersion is taken into account in this calculation. The dependence energies of calculation (solid line) and experiment (hollow circles) on the well widths are summarized in Figure 5.23(b). For calculation, the structural parameters are determined by TEM and XRD analysis [60, 63]. The observed ISBT energies are red shifted compared with the theoretical calculation, and this discrepancy increases for narrow wells. Interdiffusion at ZnSe/BeTe and CdS/ZnSe interfaces modify the QW band shape and potential profile and this effect becomes apparent in narrower wells. The observed FWHM linewidths are relatively narrow in (CdS/ZnSe)/BeTe QWs. The FWHMs increase at shorter ISBT wavelengths. As shown in Figure 5.23(c), the linewidths stand between 67–92 meV and the relative linewidth of $\Delta E/E$ at $\lambda \sim 1.55$ μm is around 10 %. We first calculated the linewidths assuming ±0.5 or ±1 ML deviation in well thickness, as given in the figure (dotted lines). This indicated that the fluctuation in the well layer is about ±0.5 ML. The huge band offset of 3.1 eV, providing

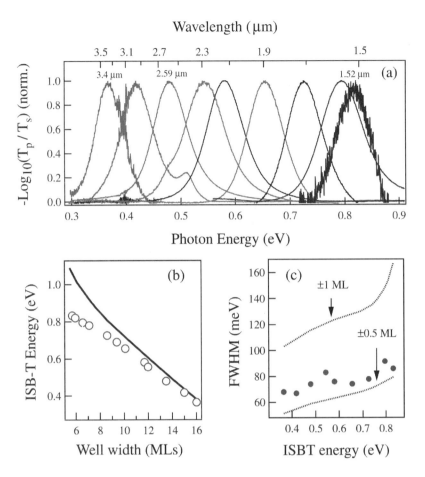

Figure 5.23 (a) Absorption spectra of (CdS/ZnSe)/BeTe QWs with different well thicknesses. The spectra have been normalized for comparison. (b) Well width dependence of e1 – e2 ISBT energies for (CdS/ZnSe)/BeTe QWs: the experimental (hollow circles), and the calculated (solid line). (c) The dependence of FWHM linewidths (filled circles) on the ISBT energies. The dotted lines indicate the calculated linewidths with assuming ±0.5 or ±1 ML fluctuation in well layer [10]. (Reproduced by permission of © 2006 American Institute of Physics (AIP))

better carrier confinement, contributes to the narrow linewidth. The dependence of linewidths on the growth temperatures (in Figure 5.22), electron doping [64] and annealing experiments indicates that carrier density and interdiffusion also affect the FWHM in this structure.

5.3.2 Waveguide Structure for a CdS/ZnSe/BeTe Gate

The latest waveguide structure [65] comprises an approximately 2.5-μm thick $Zn_{0.57}Mg_{0.28}Be_{0.15}Se$ bottom cladding layer (CL, $n \sim 2.32$–2.34), a ZnSe (10 ML)/BeTe (5 ML) MQW bottom optical confinement layer (OCL, $n \sim 2.52$–2.53), a (CdS/ZnSe)/BeTe MQW active layer (AL, $n \sim 2.52$–2.54), a ZnSe/BeTe MQW upper OCL, and an approximately 1-μm-thick $Zn_{0.57}Mg_{0.28}Be_{0.15}Se$ upper CL, all of which were sequentially grown on a

GaAs (001) substrate in a dual-chamber molecular beam epitaxy system. The corresponding refractive index n was measured at 1.55 μm using an ellipsometer for samples, each comprising a single layer only. As compared with our previous waveguides [66], the Mg and Be compositions in CL have been increased to 28 % and 15 % , respectively, in order to decrease the value of n of CL (n_{CL}), while the lattice-match condition is still maintained. The previous ZnBeSe ternary OCL ($n \sim 2.45$) has been replaced by ZnSe/BeTe MQW. By optimizing the relative thicknesses of ZnSe and BeTe, we can ensure lattice-matched growth and meanwhile increase the value of n of OCL (n_{OCL}) to as high as that of AL. As a result, the separate confinement heterostructure (SCH) layer (including an AL and both OCLs) has a nearly homogenous n, which increases the refractive index contrast between the SCH and cladding layers. Each OCL has 86 QWs and the AL has 15 QWs (~ 0.072 μm) in which each cycle comprises 1 ML ZnSe/$\sim$ 3 ML Cl-doped CdS/1 ML ZnSe as well as a 10 ML BeTe barrier. Hence, there are two main differences between the previous waveguide and the current waveguide: (i) the refractive index contrast between SCH and CL is increased, and (ii) the thickness of AL was decreased from 0.24 to 0.072 μm, i.e., a reduction of nearly two thirds. An approximately 1-μm wide high-mesa waveguide was fabricated with an approximately 3-μm wide tapered structure at the input and output edges by using a standard device process including high-resolution photolithography and reactive ion etching. The SEM image of the waveguide is shown in Figure 5.24.

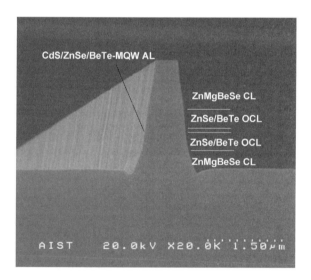

Figure 5.24   Scanning electron microscope image of cross-sectional view of high-mesa SCH waveguide

We calculate the absorption saturation curves by simulating the light propagation in the waveguide to optimize waveguide parameters and to examine the improvements in performance by new OCL. The simulation is performed by using the implicit finite-difference beam propagation method (BPM) [67]. The waveguide model is a three-dimensional tapered structure, as shown in Figure 5.25. In the simulation, AL thickness (h_{AL}), OCL thickness (h_{OCL}), refractive indices

(n_{CL}, n_{OCL}), and mesa width (w_{mesa}) are variables, while the other symbols are constants. The absorption saturation is assumed to be $\alpha = \alpha_0/(1 + I/I_{sat})$, where $\alpha(\alpha_0)$, I, and I_{sat} are the absorption coefficient, input pump intensity and saturation intensity, respectively. The simulation needs values of I_{sat} and α_0 for the AL. To obtain comparable results for different h_{AL} values, we need to know the dependence of I_{sat} and α_0 on h_{AL}. However, this is difficult to know directly. Therefore, a better approach is to set a fixed thickness with fixed parameters as a unit. Such a unit is ideally one QW layer, however, we cannot mesh waveguides into so small size in simulation. To deal with such a concept, we introduce a unit of a fixed thickness and then extrapolate to other thicknesses. The thickness unit of 0.1 µm with constant $I_{sat} = 1\,W/\mu m^2$ and $\alpha_0 = 0.081\,\mu m^{-1}$ is used to construct AL. For example, AL has three units if $h_{AL} = 0.3\,\mu m$. This method is reasonable for optimizing h_{AL} only if the carrier density is stable from one QW to another. The full transparent boundary condition, TM polarization, and the semi-vector mode are selected in simulation. The input field adopts the mode input file that is precalculated for each structure with different index profiles and the pump wavelength is 1550 nm. The computed steps along x, y, and z are 0.2, 0.01, and 0.5 µm, respectively.

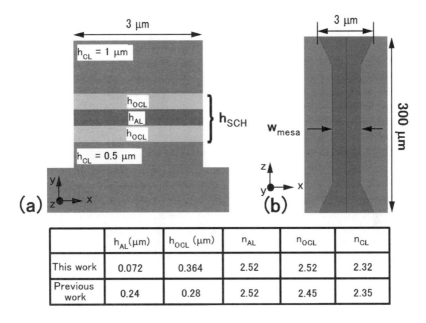

	h_{AL} (µm)	h_{OCL} (µm)	n_{AL}	n_{OCL}	n_{CL}
This work	0.072	0.364	2.52	2.52	2.32
Previous work	0.24	0.28	2.52	2.45	2.35

Figure 5.25 Schematic tapered-waveguide model: (a) cross-section view, and (b) top view with main parameters set up in BPM simulation. The inset table lists the real parameters of the waveguides in this work and our previous work (*h*: height, *w*: width) [65]. (Reproduced by permission of © 2007 Optical Society of America (OSA))

Figure 5.26(a) shows simulated absorption saturation curves for the latest [65] and previously reported waveguides [66]. Apparent improvement of saturation efficiency is observed in the latest waveguide structure due to the increase in refractive index contrast and the decrease in active layer thickness. For quantitative comparisons, the half-saturation intensity (I_{sh}) is defined in Figure 5.26(a) as the input power that decreases the absorption magnitude from the unsaturated value (T_0) to $0.5T_0$. As seen in Figure 5.26(a), the increase in refractive index contrast and the decrease in active layer thickness result in an approximately 67 % reduction in the saturation intensity. In order to separate the effects of these two changes and further to optimize the waveguide parameters, we map the half-saturation intensity against h_{OCL} and w_{mesa} for three situations with different h_{AL}, n_{CL}, and n_{OCL} values, as shown in Figures 5.26(b)–(d). In each of Figures 5.26(b)–(d), the minimum saturation intensity position at several fixed mesa widths is plotted by dots, and the broken lines between them are also shown for ease of understanding of saturation intensity behavior.

By comparing Figures 5.26(b) and (c), one can understand the role of the increase in refractive index contrast. In the waveguide with larger refractive contrast in Figure 5.26(c), the region allowing the existence of the optical mode is extended to the region having a mesa width narrower than 1 μm; thereby stronger optical intensity and lower saturation intensity will be expected. Actually, at a mesa width of less than 1 μm, one can see a slight decrease in the saturation intensity in Figure 5.26(c). However, when the saturation intensity is compared

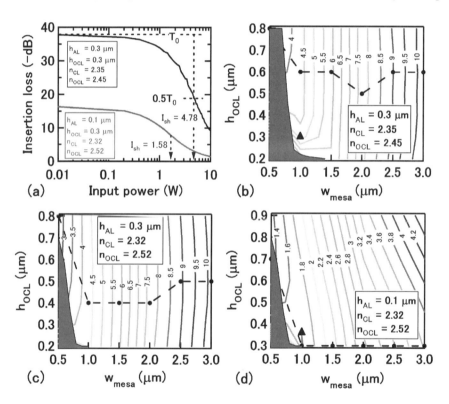

Figure 5.26 (a) Simulated absorption saturation curves for the waveguides at $w_{mesa} = 1$ μm for two waveguide parameter sets. The half-saturation intensity value (I_{sh}) of each line is shown. (b)–(d) The contour mappings of I_{sh} versus h_{OCL} and w_{mesa}. The optical mode does not exist in the black shaded regions and the dashed lines (valley paths) plot h_{OCL} at the minimum I_{sh} along w_{mesa}. The two triangles in (b) and (d) indicate the approximate positions for the previous work and this work, respectively. Minimum saturation intensity positions at several fixed mesa widths is plotted by dots, and broken lines between them are shown for ease of understanding of saturation intensity behavior [65]. (Reproduced by permission of © 2007 Optical Society of America (OSA))

at the same mesa width of more than 1 μm for both waveguide cases, no obvious difference is exhibited. On the other hand, a comparison of active layer thickness in Figures 5.26(c) and (d) reveals a >50 % decrease in saturation intensity for any values of mesa width and OCL thickness for the waveguide in Figures 5.26(d), showing the remarkable effectiveness of active layer thickness. The saturation intensity exhibits a weak dependence on the OCL thickness for mesa widths larger than 1.5 μm (Figure 5.26(b)) and 1 μm (Figure 5.26 (c)), when thicker AL ($h_{AL} = 0.3$ μm) is compared. In contrast, in the case of thinner AL ($h_{AL} = 0.1$ μm), a thinner OCL thickness results in an apparent decrease in the saturation intensity for each mesa width, as shown in Figure 5.26(d). In all these three figures, the saturation intensity always decreases with decreasing the mesa width for any OCL thickness due to the improved optical confinement. Therefore, we can draw the following conclusions as a guide for waveguide design: (i) decreasing the AL thickness can significantly decrease the saturation energy; (ii) decreasing OCL thickness is more effective for thinner Als, and (iii) the lowest saturation intensity occurs close to the edge of the mode-inhibited region that depends on the refractive index contrast and mesa width. The numerical results indicate lower saturation energy in our current waveguide with thinner AL, which is confirmed in the following sections.

5.3.3 Characteristics of a CdS/ZnSe/BeTe Gate

Figure 5.27(a) shows the TM and transverse-electric (TE) insertion loss versus the input pulse energy at a control wavelength of 1565 nm that is nearly resonant with the ISBT energy (~1.57 μm) determined by Fourier-transformed infrared spectroscopy. The pulse duration is found to be ~0.34 ps using an autocorrelator. Absorption saturation obviously occurs for TM polarization with increasing input pulse energy, while the TE insertion loss remains at a relatively low level (<6 dB). For quantitative comparison, the 10-dB saturation energy is defined as the input energy that can induce a 10 dB increase in transmittance with respect to the unsaturated insertion loss. As shown in Figure 5.27(a), the 10 dB saturation energy at 1565 nm is about 2 pJ. In contrast to our previous results with a comparable mesa width [66], we achieved a reduction of nearly 70 % in the 10-dB saturation energy (decreasing from 7 to

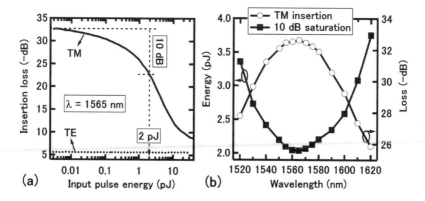

Figure 5.27   (a) The waveguide absorption saturation curves of TM polarization at 1565 nm with a definition of 10 dB saturation energy. (b) Wavelength-dependent 10 dB saturation energy and TM insertion loss [65]. (Reproduced by permission of © 2007 Optical Society of America (OSA))

2 pJ). Based on the above numerical results, this remarkable reduction can be mainly attributed to the decrease in the thickness of AL. Considering the real application, we measured such absorption saturation curves at different wavelengths from 1520–1620 nm and then obtained the wavelength-dependent 10-dB saturation energy, as shown in Figure 5.27(b). The figure also shows the unsaturated TM insertion loss versus the wavelength that exhibits the absorption spectral characteristics. Both curves have peak positions at about 1565 nm, indicating that the saturation energy reaches a minimum value when it is resonantly pumped. When the photon energy deviates from the resonant energy, the 10-dB saturation energy gradually increases, due to the decrease in pump efficiency, while it still maintains a very low level (2.0–2.2 pJ) for 1.55–1.58 μm. This is beneficial for future device designs.

Such low saturation energies are expected to lead to correspondingly low values in switching energy (E_s) in ultrafast switching operations. To confirm this expectation, the pump-energy-dependent transmission of the signal was measured by a time-resolved, two-color pump-probe experiment with a 1520-nm beam as the control pulse, and a 1566-nm beam as the signal pulse, both of which were aligned with TM polarization. Their pulse durations were about 0.34 ps and about 0.44 ps, respectively and the probe pulse energy was as low as 0.1 pJ. Figure 5.28 illustrates the temporal signal transmittance for different control pulse energies. When two pulses coincide, the signal transmittance is obviously enhanced. This can be clearly illustrated by plotting the maximum increase in signal transmittance (i.e., switching extinction ratio) against the corresponding pump energy, as shown in the inset of Figure 5.28. With an increase in the control pulse energy, the extinction ratio first increases linearly and then exhibits saturation characteristics. As seen, < 7 pJ pump energy at 1520 nm induces the 10-dB increase in probe transmission, which is currently the lowest reported E_s (7 pJ/10 dB) in ISBT switches. A further

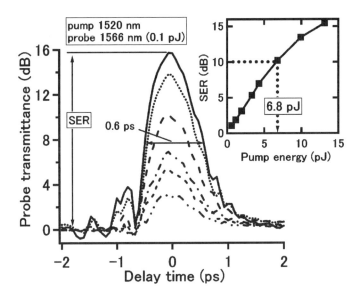

Figure 5.28 Temporal transmitted probe intensity versus pump-probe delay-time under different pump energies (increase along *y* axis). The inset shows the pump-energy dependent SER [65]. (Reproduced by permission of © 2007 Optical Society of America (OSA))

reduction of about 38 % is obtained as compared with our previous value of 11.3 pJ/10dB [66]. Furthermore, the change in the signal transmittance returns to almost zero within ~1–2 ps. Judging from the 0.6 ps full width at half maximum of the absorption saturation recovery, the waveguide can work at bit rates of the order of 800 Gbps–1 Tbps.

The absorption saturation is influenced by an ISBT broadening mechanism that is related to various intersub-band relaxation processes [68]. So we briefly discuss the potential homogeneous and inhomogeneous origins in our system ($E_{ISBT} > \hbar\omega_{LO}$). Well-to-well thickness fluctuations and sub-band nonparabolicity are fundamental inhomogeneous broadening mechanisms. For our sample growth, we did not observe the well-thickness fluctuation-induced peak separation even when the well thickness was systematically increased. Compared to III–V, this system has a weaker sub-band nonparabolicity based on the energy-dependence effective mass [69]. Furthermore, the sheet density is about $10^{11}–10^{12}$ cm^{-2} in our sample, and the many-body effects will cancel the effect of sub-band nonparabolicity on inhomogeneous broadening [68]. So homogeneous broadening may be the main broadening mechanism, even though the shape of the absorption spectrum (300 K) is not a pure Lorentzian line. The decay time constant is about 0.2 ps, mainly due to LO phonon scattering, as referred to in Ref. [70]. Carrier relaxation in our current quantum system is free from the influence of the X valley of BeTe, which is different from the observed carrier-relaxation mediation of indirect valleys [58].

5.4 InGaAs/AlAs/AlAsSb ISBT Gate

The InGaAs/AlAs/AlAsSb ISBT exhibits the longest relaxation time, of the order of ps, when compared with other material system quantum wells. Then we can expect a lower saturation energy; however, the saturation energy obtained so far was still unsatisfactory. This is mainly due to the large two-photon absorption (TPA) because of the rather narrow bandgap of this system. In this section we describe the present state of the device structure and characteristics as a saturation type gate. In Section 5.5 we will describe the newly found cross-phase modulation effect in this device, which is highly promising for overcoming the drawbacks of this material system device.

5.4.1 Device Structure and its Fabrication

The device structure and the typical quantum well structure are shown in Figure 5.29 [13,71]. We employed a coupled double quantum well (C-DQW) structure as shown in Figure 5.29(b). This is because the well depth of this sample is rather shallow as compared with other material systems [72]. The energy splitting due to the coupling of quantum wells gives a transition wavelength of 1.55 µm for the ground state to fourth state transition. In addition to this, the C-DQW structure gives a faster response, due to additional electron relaxation paths [14]. The C-DQW structure was grown by MBE. In the coupled double quantum-well structure, we inserted a few monolayers of AlAs between the InGaAs well layer and the AlAsSb barrier layer. This was to prevent interfacial diffusion of the composite atoms associated with heavy doping of Si ($1–2 \times 10^{19}$/cm^3) [73]. The In composition of the well layer is increased to 0.8 to give a larger lattice constant to compensate for the strain due to the center barrier layer AlAs and interfacial AlAs layer, which has a smaller lattice constant. The well layer thickness is 2.3–2.7 nm, the center barrier AlAs layer is 4 monolayers. The quantum wells are stacked to form a waveguide as shown in Figure 5.29. The larger optical confinement is desirable for low switching energy operation. The refractive index of AlAs and AlAsSb is 2.89 and 3.05 respectively; they are comparatively small and decrease the optical confinement factor of the waveguide. So the structure is designed to minimize the thickness of these layers. The interfacial AlAs layer was designed to be 2 ML, and the AlAsSb layer was reduced to 2 nm, which is the limit to prevent the formation of a miniband through coupling between C-DQWs. The doping is done only to the central region of the stacked C-DQWs using Si (1×10^{19}/cm^3), and undoped C-DQWs at upper and lower stacks serve as waveguiding layers. Around 17 pairs of C-DQWs are doped at the center region, and 25 to 30 pairs of upper and lower layers of C-DQW are undoped. An example of an absorption spectrum measured for a wafer is shown in Figure 5.30. The peak close to 1.55 μm corresponds to the ground state to fourth sub-band transition, and the larger peak is second-to-third level transition. The larger absorption coefficient for the 2 to 3 transition is due to the form of the envelope wave function. We use the absorption saturation of 1-to-4 transition. The lateral waveguide is a narrow mesa structure with a width of about 1 μm, and tapered waveguides are attached to both ends as in the GaN/AlN, and the CdS/ZnSe/BeTe ISBT devices. The optical confinement factor for the doped quantum wells is calculated at around 0.3. The mesa waveguide is tilted by 7 degrees to reduce the facet reflection, and anti-reflection coating was done to the facets. The device was coupled with a polarization-maintaining optical fiber to form a pigtailed module. The characteristics were evaluated for the module. The insertion loss was measured for TE light, which is not absorbed by ISBT. The insertion loss including the coupling loss to both fibers was 3–5.5 dB.

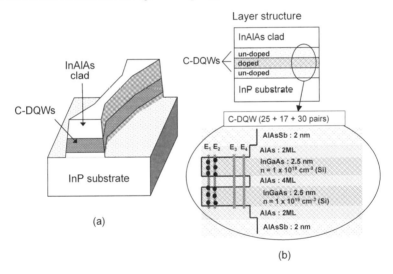

Figure 5.29 Device structure (a), and waveguide and quantum well structure (b) of InGaAs/AlAs/AlAsSb ISBT gate device [71]. (Reproduced by permission of © 2007 Japan Society of Applied Physics (JSAP))

5.4.2 Saturation Characteristics and Time Response

Figure 5.31 shows the absorption saturation characteristics of the module. The vertical axis is the transmittance, and the horizontal axis is the fiber input energy/pulse. The solid line is the

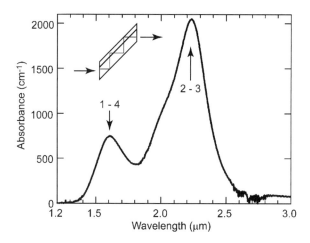

Figure 5.30 An example of absorption spectrum. Absorption saturation of a 1–4 transition near 1.55 μm is used for gate operation [13]. (Reproduced by permission of © 2007 IEEE)

transmittance for the TM polarized control pulse for a width of 340 fs. As an increase of the control energy, transmittance increases showing the absorption saturation. However at high energy, transmittance starts to saturate and decreases. This is due to the TPA. In the figure is also plotted the transmittance for TE light (dotted line). This decreases with increase of control pulse energy, reflecting the TPA. The narrow badgap of this system results in the large TPA. The plots by closed circles are for the probe pulse about 1 ps after the control pulse. In this case, a control pulse of width 150 fs was used. Larger extinction ratios were observed for the delayed probe pulse. This is due to the reduced TPA originating from the cross term of control

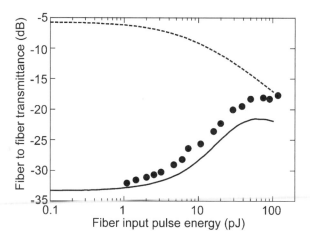

Figure 5.31 Absorption saturation characteristics of InGaAs/AlAs/AlAsSb ISBT gate module. The dotted line is the transmission for TE mode, the bold line is the transmission for TM control pulse for the control pulse width of 340 fs. The plots by closed circle are transmission for TM probe pulse delayed 1 ps after the control pulse of the width of 150 fs [13]. (Reproduced by permission of © 2007 IEEE)

pulse and probe pulse. As can be seen from these results, the TAP is one of the crucial factors for realizing lower energy operation.

Figure 5.32 shows the time response of the gate. The transmittance for the probe pulse is shown as a function of time. The control pulse was injected at the time indicated by the arrow. The peak transmittance is obtained about 1 ps after the control pulse where the TPA due to the cross term of control and probe vanishes. We can see the response time of the order of ps. We also see the dependence of recovery time on the pump energy. The recovery time is faster for a higher pump energy but the reason for this is yet to be established. It is probable that higher energy excitation causes faster energy relaxation of electrons due to an enhanced scattering rate. For a high pump energy of 118 pJ, we can get an extinction ratio of 15.6 dB. However, the pulse energy necessary to get the extinction is too high for practical use. This is a problem to be overcome to make this device useful.

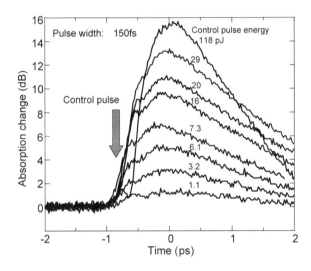

Figure 5.32 Time response of InGaAs/AlAs/AlAsSb ISBT gate. Control TM pulse with width of 150 fs is injected at the arrowed position. Then the transmittance for the probe pulse is measured. A control pulse with energy of 118 pJ provided an extinction ratio of 15.6 dB

5.5 Cross-phase Modulation in an InGaAs/AlAs/AlAsSb-based ISBT Gate

Previous sections have described all optical gates that are based on the absorption saturation of intersub-band transition (ISBT) in semiconductor quantum wells (QWs). Although subpicosecond switching operations have been demonstrated in these devices, repetitive operation beyond 10 Gbit/s has not been achieved, mainly due to the relatively high switching energy of the order of several tens pJ. This section presents a novel operation mode for InGaAs/AlAs/AlAsSb ISBT devices, in which ISBT absorption of a transverse-magnetic (TM) light gives rise to ultrafast cross-phase modulation (XPM) of a co-propagating, transverse-electric (TE) light [15, 74]. An important point is that the phase modulation takes place in lossless TE probe light. This enables us to realize ultrafast, low-insertion loss, all-optical devices for various applications. In addition, this phase modulation effect has very wide bandwidth for the signal light, and thus is very useful for signal processing.

First we describe this phase modulation effect and then discuss the possible mechanisms. As an application we demonstrate an all-optical wavelength conversion for 10-Gbit/s return-to-zero (RZ) signals with pulse width of 2.6 ps to demonstrate the feasibility of this effect for all-optical signal processing devices.

5.5.1 Cross-phase Modulation Effect and its Mechanisms

The new phase modulation effect was found when we put in the TE polarized CW probe light to the ISBT gate switch module and pumped using TM polarized control light with repetition of 10 GHz and a pulse width of 2.6 ps. Figure 5.33 illustrates the spectrum of TE probe light showing the phase modulation. The intensity of the pump pulse to the fiber was 8 pJ. We can see sidebands showing the phase modulation of TE probe light. As will be shown in Section 5.5.2, refractive index change corresponding to this phase shift is to the positive side and its response speed is very fast.

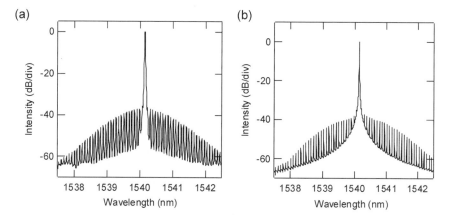

Figure 5.33   (a) Phase modulation spectrum of the CW probe light when the TM control (10 GHz, 1.7 ps pulse width) was put into the ISBT gate module. (b) Calculated power spectrum of the phase modulation. Good fit with the experimental result was obtained for the peak phase shift of 0.35 rad

The amount of phase shift can be estimated by comparing the experimental spectrum and power spectrum obtained by Fourier transform of the phase modulated electric field:

$$E(t) = E_0 \exp[i\{\omega t + \Delta\varphi(t)\}], \qquad \Delta\varphi(t) = \Delta\phi_0 \sum_{m=-\infty}^{\infty} \exp\left(-\frac{(t - mt_0)^2}{t_w^2}\right) \qquad (5.13)$$

where $\Delta\varphi_0$ is peak phase shift, t_0 is the period of 10 GHz, $2\sqrt{\ln 2}\, t_w$ is the pulse width, and m is an integer. Figure 5.33(b) shows the calculated spectrum where amplitude of phase modulation is adjusted to match the measured spectrum. Good agreement was obtained for a maximum phase shift of 0.35 rad. In the first report, we estimated the amount of phase shift by changing the phase modulation to the amplitude modulation using a delayed interferometer [74]. In this experiment we used the EDFA. However, it turned out that the saturation of the EDFA tended to overestimate the amount of phase shift. The typical amount of phase shift was of the order of 0.02–0.05 rad/pJ. It is difficult to get a precise estimate of the refractive index change associated with this phase shift, because the TM pump power decreases steeply as it propagates in the device. The pump power is mostly absorbed within 100–200 μm because of the a few hundred cm^{-1} units of absorption coefficient. As a rough estimate, if we assume the phase shift distance to be 150 μm and the optical confinement factor to the quantum well layer to be 0.1, the refractive index change for the phase shift of 0.5 rad is 8.2×10^{-3}. This is a very large refractive index change.

Phase modulation of the TE probe light by TM pumping is a very strange phenomenon. The ISBT takes place only for the TM light. The TM pumping causes absorption saturation for TM light and then it causes a refractive index change for TM light as predicted by the Kramers–Kronig relationship. However, there is no reason for the phase modulation in TE light. The dipole moment of ISBT is only for TM mode. What is interesting is that no phase shift in the CdS/ZnSe/BeTe ISBT switch module was observed. We have considered two possible mechanisms: one is the plasma dispersion effect associated with the nonparabolicity of the conduction band, and the other is dispersion due to the interband transition.

Firstly we discuss the plasma dispersion effect [15]. Figure 5.34 shows a diagram of the conduction band in coupled double quantum-wells (C-DQWs) along with calculated energy levels, quasi-Fermi level for 1.7×10^{19}/cm^3 carrier density (sheet carrier density of 8.8×10^{12}/cm^2), Γ-point effective masses of electrons, and intersub-band relaxation times. In the calculation, well depth was put at 1.4 eV, despite the experimentally estimated value of 1.6 eV for a wide quantum well [72]. This was because we assumed an interfacial diffusion that deforms the quantum well in our very thin quantum wells. In the calculation of energy levels, we used the band nonparabolicity of a bulk semiconductor as expressed by:

$$\frac{1}{m^*} = \frac{1}{m_0} + \frac{2P^2}{3m_0} \left(\frac{2}{E_g + E} + \frac{1}{E_g + E + \Delta} \right) \qquad (5.14)$$

where m_0 is the electron mass, P is the momentum matrix element, E_g is the band gap of the well layer, E is the energy of each level, and Δ is the energy difference between the valance band edge and the split-off band. Although the exact calculation requires a $\mathbf{k} \cdot \mathbf{p}$ perturbation calculation for C-DWQs including strain and anisotropy of the band [75–77], we used the simple expression shown above and parameters for bulk In$_{0.8}$Ga$_{0.2}$As in order to make a rough estimate the refractive index change. It can be seen from Figure 5.34 that the first and second sub-bands are located below the Fermi level at the Γ-point and that the effective masses are heavier at higher energy sub-bands.

As a TM control light we assumed a Gaussian pulse with a full width at half maximum (FWHM) of 1 ps, and calculated the evolution of electron density using rate equations. The transition time constants were assumed as indicated in Figure 5.34. The intensity of the TM control pulse was adjusted so as to induce an absorption coefficient reduction of 50 %, corresponding to a 14-dB increase in the transmittance of our typical devices. Figure 5.35 shows the calculated time evolution of the electron density for each sub-band. It can be seen that the electron density decreases in the first and second sub-bands, while it increases in the third and fourth sub-bands.

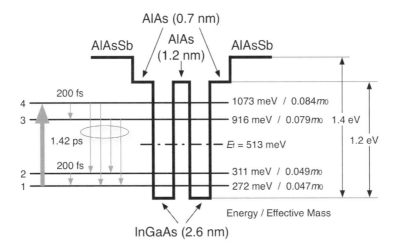

Figure 5.34 Schematic diagram of the conduction band in C-DQWs along with calculated energy levels, quasi-Fermi level, Γ-point effective masses of electrons, and intersub-band relaxation times

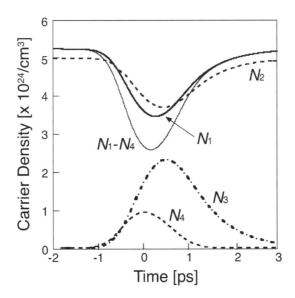

Figure 5.35 Calculated results for the time evolution of the carrier density in each sub-band caused by the absorption of a 1-ps TM pump pulse [15]. (Reproduced by permission of © 2007 Institute of Pure and Applied Physics (IPAP))

We assumed the plasma confined in the quantum well to be an ideal two-dimensional plasma for purposes of analysis, and expressed the plasma frequency as [78]:

$$\omega_{\mathrm{pl}}(q) = \sqrt{\frac{e^2 N_{\mathrm{s}}}{2\varepsilon_0 m^*}q} \qquad (5.15)$$

where q is the wave number of photon, e is the electron charge, N_s is the sheet electron density, and ε_0 is the vacuum permittivity. The refractive index change due to plasma effects can be represented by Drude's formula and is obtained by summing up the contribution from each sub-band:

$$\Delta n = -\sum_{i=1}^{4} \frac{e^2 N_{s,i} q}{4\varepsilon_0 m_i^* \omega^2}. \tag{5.16}$$

Here the suffix i denotes the sub-band number and ω represents the angular frequency of the probe light. When the effective masses of all the four sub-bands are equal, the summation in Equation (5.16) remains the same for the change in sheet electron density at each sub-band by TM pulse excitation. However, due to the difference in the effective masses as shown in Figure 5.34, optical pumping results in a refractive index change, which is clearly displayed in the calculated results in Figure 5.36. Corresponding to the carrier density change, the dispersion in the first and second sub-bands increases the refractive indices, while that in the third and fourth sub-bands decreases the refractive indices. The overall effect is an increase in refractive index of the order of 10^{-3}. If the effective masses of the sub-bands are equal, there occurs no refractive index change.

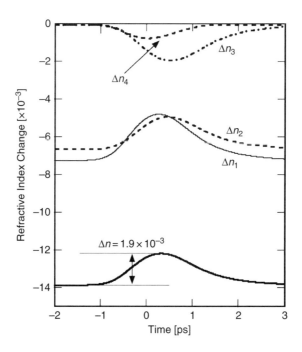

Figure 5.36 Calculated results for the time evolution of the refractive index change caused by the absorption of a 1-ps TM pump pulse, where Δn_i represents the contribution from the ith sub-band [15]. (Reproduced by permission of © 2007 Institute of Pure and Applied Physics (IPAP))

Another possible cause of phase modulation is the dispersion resulting from interband absorption [2,14]. Depletion of the electrons in the ground state in the conduction band caused by TM control light absorption allows interband transition from the valence band to conduction band. In compressively strained quantum wells, the topmost valence band is a heavy-hole-like band and transition takes place for TE mode. If we assume a newly generated Lorenzian shaped interband absorption spectrum, the corresponding refractve index is given by:

$$n = \int_0^\infty \frac{1}{\Delta E \sqrt{\pi}} e^{-\frac{(E_{ib}-E_0)^2}{\Delta E^2}} \left(1 + \frac{\alpha c n_0 \hbar^2 \gamma (E_{ib} - E)}{2E\left[(E_{ib} - E)^2 + \hbar^2 \gamma^2\right]}\right) dE_{ib} \qquad (5.17)$$

where we assume Gaussian inhomogenous broadening and $2\Delta E$ corresponds to the e^{-1} width of inhomogenous broadening. E_{ib} is the interband transition energy, α is the absorption coefficient, n_0 is the original refractive index, and γ is the dephasing time. Figure 5.37 shows the calculated refractive index change, assuming an inhomogenous broadening of 40 meV, a dephasing time of 200 fs, an original refractive index of 3.3, and a newly generated peak absorption coefficient of 100–500 cm^{-1}. The absorption peak energy is assumed to be 1.0 eV. As can be seen from Figure 5.37, the corresponding refractive index change at a wavelength of 1.550 nm (0.8 eV) is of the order of 10^{-4}. However, we can see that if we design a quantum well whose band-to-band transiton energy is closer to 1.55 μm, the interband effect becomes greater.

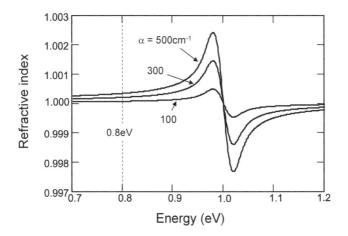

Figure 5.37 Calculated refractive index originating from newly generated interband absorption

In the actual device, both effects may be taking place. At present the models are rough ones. In order to obtain a more quantitative estimate, including percentages of the contribution of plasma dispersion effect and that of band edge effect, a more rigid analysis is needed with evaluation of the dependence of phase shift on the probe wavelength. One attractive aspect is that if we could design a quantum well where the band-edge effect dominates without increasing the band-to-band absorption, we can expect polarization-independent refractive index changes, because both heavy hole and light hole bands can contribute to interband absorption.

It is interesting to note that XPM has not been observed in CdS/ZnSe-based ISBT devices [15]. This can be explained for both mechanisms. As for nonparabolicity-related plasma dispersion, the large effective $0.2 m_0$ in CdS makes the plasma dispersion effect negligible. The plasma frequency of InGaAs corresponds to a wavelength of 12.5 μm, while that of CdS/ZnSe is 29 μm for a sheet carrier density of $5.31 \times 10^{12}/cm^2$ at the ground state, and the wider band gap of CdS/ZnSe makes band nonparabolicity smaller, which results in a refractive index change below 1×10^{-4} at 1550 nm. It is obvious that the band edge effect can be ignored in CdS/ZnSe/BeTe ISBT devices, because of the very wide band gap of 2.48 eV.

5.5.2 Application to Wavelength Conversion

As an application of this phase modulation effect, we performed a wavelength conversion experiment in which the phase modulation in the CW TE probe light, whose wavelength differs from that of the TM control light, was converted to the intensity modulation, and then the error rate was evaluated. Devices used in the experiments had InGaAs/AlAs/AlAsSb C-DQWs as described in Section 5.3. Two samples were used in the experiments, each mounted on pig-tailed modules with a thermo-electric cooler. Sample No.1 was a 360-μm long mesa waveguide device with mesa widths of 1 μm, while sample No.2 had a 240-μm long ridge waveguide structure with a ridge width of 2 μm. Insertion losses for a TE probe light were 4.66 and 3.72 dB for the samples 1 and 2, respectively.

Figure 5.38 shows a diagram of the experimental set-up [74]. A continuous-wave (CW) light at 1559.9 nm was used as a TE probe signal. An actively mode-locked fiber laser (AML-FL) with a center wavelength of 1550.1 nm was employed as a TM control light source. A 9.95328-Gbit/s return-to-zero (RZ) data stream was generated from the AML-FL by encoding its output pulses by a $2^{31}-1$ psuedorandom bit sequence (PRBS) pattern with a LiNbO$_3$ intensity modulator. Inset (a) in Figure 5.38 represents the eye diagram of the modulated pump signal with a pulse width of 2.3 ps. For efficient coupling of pump power to the sample, a counter-propagating configuration was employed, in which the TE probe and TM control lights were input from the right and left sides of the module, respectively. Since ISBT absorption occurs only for a TM wave, the TE probe light propagated through the sample without attenuation excluding a coupling loss of about 3 dB. For the conversion of phase modulation to intensity

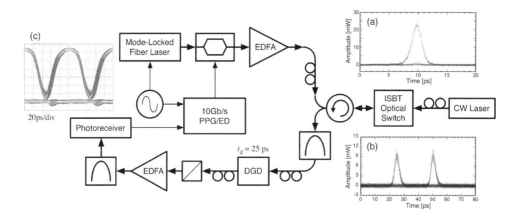

Figure 5.38 Diagram of the experimental setup for 10-Gb/s wavelength conversion [74]. (Reproduced by permission of © 2007 Optical Society of America (OSA))

modulation, we used a polarization interferometer, which consisted of a differential group delay (DGD) generator, polarizer, and polarization controllers. The DGD was 25 ps. In the experiments, the phase bias of the polarization interferometer was set at π radians, which produced no output signal in the absence of the control light. When we put in the TM control light, we obtained an output consisting of two pulses with a time difference corresponding to the DGD of 25 ps. The optical sampling trace of the probe light at the interferometer output is shown by inset (b) in Figure 5.38, which contains 2-bit RZ signals with 25 ps separation and shows clear opening with an extinction ratio of 33 dB. The temporal width of these signals is 5.0 ps, which is about twice that of the pump pulse. This broadening is due to the counter-propagating configuration of the control and probe lights, where the propagation time through the sample is 3.9 ps. It should be noticed that the shape of these XPM-induced signal is almost symmetrical, indicating that XPM has short rising and falling times. This signal was detected by a 35-GHz photoreceiver followed by a 12.5-GHz limiting amplifier. Inset (c) in Figure 5.38 shows the electrical sampling trace of the limiting amplifier output, in which the 2-bit signals were converted into a single-bit signal due to the limited bandwidth of the amplifier. The output signal from the amplifier was fed to an error detector (ED) for bit error rate (BER) measurement.

Figures 5.39(a) and (b) show the optical sampling traces and optical spectra of the probe light at the interferometer output, respectively, measured by changing the control pulse energy. The measured BER is shown in Figure 5.40 as a function of the received optical power. Curve A represents the result for back-to-back operation and the receiver sensitivity was -19.9 dBm at BER $= 10^{-9}$. Curves B and C were obtained with the samples 1 (pump energy 6.62 pJ) and 2 (7.10 pJ), respectively. Error free operation was achieved with a power penalty of 5.3 and 2.5 dB for samples 1 and 2, respectively.

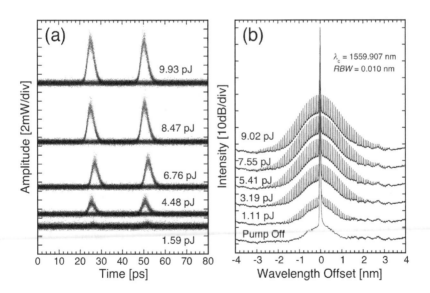

Figure 5.39   (a) Optical sampling traces and (b) optical spectra of the probe light at the interferometer output measured by changing the pump pulse energy [74]. (Reproduced by permission of © 2007 Optical Society of America (OSA))

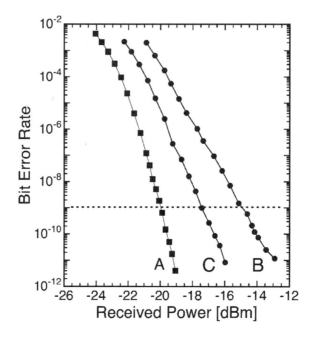

Figure 5.40 BER curves plotted as a function of the received power. Curve A corresponds to back-to-back operation, while curves B and C are obtained with the samples 1 and 2, respectively [74]. (Reproduced by permission of © 2007 Optical Society of America (OSA))

From the practical viewpoint, the polarization interferometer used in the experiment is not desirable, because it generates 2-bit signals from 1-bit input and cannot be used for gating clock signals. These drawbacks can be overcome by employing a gate switch configuration as shown in Figure 5.41(a). An ISBT device was placed in the upper arm of an asymmetric Mach–Zehnder interferometer, while a variable optical attenuator (VOA) and a phase shifter were placed in the lower arm for amplitude and phase adjustment, respectively. For achieving stable

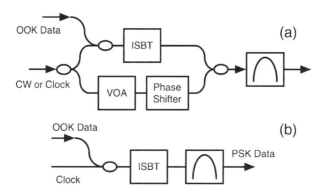

Figure 5.41 Schematic diagrams of (a) optical gate switch and (b) OOK-PSK modulation format converter [74]. (Reproduced by permission of © 2007 Optical Society of America (OSA))

operation, monolithic- or hybrid-integrated devices are desirable. This gate-switch device can be used not only as a wavelength converter but also as a demultiplexer for optical time-division multiplexed (OTDM) signals. A prototype Mach–Zehnder interferometer-type module was recently developed using space optics and error free demultiplexing operation from 160 Gb/s to 10 Gb/s was demonstrated [79]. Due to intrinsic fast response of XPM, ISBT devices can be employed for ultrafast phase modulation. Figure 5.41(b) shows diagrams of a modulation format converter that produces a binary phase shift keying (BPSK) signal from an on/off keying (OOK) RZ signal. For realizing this function, the ISBT device should produce a phase shift of π radians with a reasonable pump power. The design of the quantum well for larger phase shift is a key to this end.

5.6 Summary

The operating principle of the ISBT all-optical gate switch is presented and state-of-the-art developments such as absorption saturation type gates are presented for three different material systems. A promising new phenomenon of all-optical, cross-phase modulation effects is also presented for InGaAs/AlAs/AlAsSb devices.

As absorption saturation type gates, it is interesting to plot the relationship between the absorption saturation energy and the response time as shown in Figure 5.42. The optical pulse energy to the fiber needed to obtain an extinction of 10 dB is plotted as a function of response time.

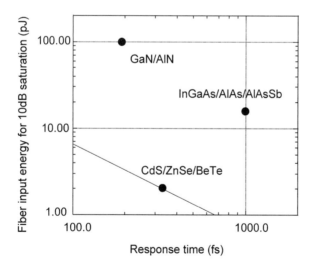

Figure 5.42 Fiber input pulse energy for 10 dB extinction versus response time of the ISBT module for three material systems

For GaN/AlN, a measured response time of 190 fs was used. For CdS/ZnSe/BeTe devices we assumed it to be 340 fs (equal to the pulse width of measurement) because direct evaluation of the response time is yet to be done. For InGaAs/AlAs/AlAsSb, a typical time constant of 1 ps was used. As the saturation energy is inversely proportional to response time as can be seen from Equation (5.7), a straight line is drawn starting from the top of the CdS/ZnSe/BeTe data point. The GaN/AlN gate shows higher saturation energy than the line. This is mainly due to the edge dislocation that causes excess loss to TM polarized light, and yet smaller optical confinement compared with that of the CdS/ZnSe/BeTe gate. The InGaAs/AAs/AlAsSb gate shows rather high saturation energy despite its slower response time. This is due to the large TPA and small optical confinement caused by the relatively low refractive index of the C-DQW layers. The lowest energy attained in the CdS/ZnSe/BeTe device was due to its high crystal quality, no TPA, and very large optical confinement to the well layer affordable in this material system.

It should be noted that the InGaAs/AlAs/AlAsSb device shows the lowest saturation energy for wider optical pulse, for example a pulse width of 2.5 ps, which is the RZ pulse width for 160 Gb/s. This is due to the larger τ_{21} and τ_{ph}. For a pulse width wider than the dephasing time, the saturation energy increases proportionally to the pulse width, then the GaN/AlN, and CdS/ZnSe/BeTe devices, having very small dephasing times, show a large saturation power for a very wide pulse width. Then for practical 160-Gb/s system applications, the InGaAs/QAlAs/AlAsSb system is promising. However, the saturation energy is still too high to use as a gate switch due to the TPA and insufficient optical confinement. The GaN/AlN device shows the fastest response time of 190 fs. High pulse energy is needed because of the very fast response, but this could be acceptable due to the material's robustness for heating. The CdS/ZnSe/BeTe device shows a very fast and at the same time fairly low switching energy of 2 pJ for 10-dB extinction. GaN/AlN and CdS/ZnSe/BeTe devices are suited for much higher bit-rates of around 1 Tb/s. Devices of these three different material systems still need further improvement. Reduction of the edge dislocation is needed in the GaN/AlN device, and further increase of the optical confinement will lead to a still lower saturation energy in CdS/ZnSe/BeTe devices, and development of some means to reduce the effect of TPA and increase the optical confinement are necessary in InGaAs/AlAs/AlAsSb devices.

A very interesting new phenomenon of cross-phase modulation is presented. The phase modulation takes place in the TE probe light, which is not absorbed by ISBT, by TM control pulse. This enables us to realize low-insertion loss, all-optical signal processing devices, overcoming the disadvantages of the large insertion loss of absorption saturation type devices. We discussed possible mechanisms of phase modulation and demonstrated the wavelength conversion for 10-Gb/s signals with a pulse width of 2.6 ps. Quite recently, a Mach–Zehnder type switch module was developed using space optics, and error free DEMUX operation from 160 Gb/s to 10 Gb/s was demonstrated [79]. This new cross-phase modulation effect is highly promising for ultrafast signal processing.

References

[1] L. C. West and S. J. Eglash, 'First observation of an extremely large-dipole infrared transition within the conduction band of a GaAs quantum well,' *Appl. Phys. Lett.*, **46**, 1156–1158 (1985).

[2] S. Noda, T. Uemura, T. Yamashita, and A. Sasaki, 'All-optical modulation using an n-doped quantum-well structure,' *J. Appl. Phys.*, **68**, 6529–6531 (1990).

[3] B. K. Ridley, 'Electron-phonon interactions in 2D systems,' in J. Shah (Ed.) *Hot Carriers in Semiconductor Nanostructures: Physics and Applications*, San Diego: Academic Press, pp. 15–51(1992).

[4] C. Gmachl, H. M. Ng, S.-N. G. Chu, and A. Y. Cho, 'Intersub-band absorption at $\lambda \sim 1.55\,\mu m$ in well- and modulation-doped GaN/AlGaN multiple quantum wells with superlattice barriers,' *Appl. Phys. Lett.*, **77**, 3722–3724 (2000).

[5] K. Kishino, A. Kikuchi, H. Kanazawa, and T. Tachibana, 'Intersub-band transition in $(GaN)_m/(AlN)_n$ superlattices in the wavelength range from 1.08 to 1.61 µm,' *Appl. Phys. Lett.*, **81**, 1234–1236 (2002).

[6] N. Iizuka, K. Kaneko, and N. Suzuki, 'Near-infrared intersub-band absorption in GaN/AlN quantum wells grown by molecular beam epitaxy,' *Appl. Phys. Lett.*, **81**, 1803–1805 (2002).

[7] D. Hofstetter, S.-S. Schad, H. Wu, W. J. Schaff, and L. F. Eastman, 'GaN/AlN-based quantum-well infrared photodetector for 1.55 µm,' *Appl. Phys. Lett.*, **83**, 572–574 (2003).

[8] A. Helman, M. Tchernycheva, A. Lusson, E. Warde, F. H. Julien, Kh. Moumanis, G. Fishman, E. Monroy, B. Daudin, D. Le Si Dang, E. Bellet-Amalric, and D. Jalabert, 'Intersub-band spectroscopy of doped and undoped GaN/AlN quantum wells grown by molecular-beam epitaxy,' *Appl. Phys. Lett.*, **83**, 5196–5198 (2003).

[9] R. Akimoto, K. Akita, F. Sasaki, and T. Hasama, 'Sub-picosecond electron relaxation of near-infrared intersub-band transitions in n-doped (CdS/ZnSe)/BeTe quantum wells,' *Appl. Phys. Lett.*, **81**, 2998–3000 (2002).

[10] B. S. Li, R. Akimoto, K. Akita, and T. Hasama, '$\lambda \sim 1.49-3.4$ µm intersub-band absorptions in (CdS/ZnSe)/BeTe quantum wells grown by molecular beam epitaxy,' *Appl. Phys. Lett.*, **88**, 221915 (2006).

[11] B. Sung, H. C. Chui, M. M. Fejer, and J. S. Harris, Jr., 'Near-infrared wavelength intersub-band transitions in high indium content InGaAs/AlAs quantum wells grown on GaAs,' *Electron. Lett.*, **33**, 818–820 (1997).

[12] T. Akiyama, N. Georgiev, T. Mozume, H. Yoshida, A. V. Gopal, and O. Wada, 'Nonlinearity and recovery time of 1.55 µm intersub-band absorption in InGaAs/AlAs/AlAsSb coupled quantum wells,' *Electron. Lett.*, **37**, 129–130 (2001).

[13] T. Simoyama, S. Sekiguchi, H. Yoshida, J. Kasai, T. Mozume, and H. Ishikawa, 'Absorption dynamics in all-optical switch based on intersub-band transition in InGaAs-AlAs-AlAsSb coupled quantum wells,' *IEEE Photonics Technol. Lett.*, **19**, 604–606 (2007).

[14] H. Yoshida, T. Simoyama, A. V. Gopal, J. Kasai, T. Mozume, and H. Ishikawa, 'Ultrafast all-optical switching and modulation using intersub-band transitions in coupled quantum well structures,' *IEICE Trans. Electron.*, **E87-C**, 1134–1141 (2004).

[15] H. Ishikawa, H. Tsuchida, K. S. Abedin, T. Simoyama, T. Mozume, M. Nagase, R. Akimoto, T. Miyazaki, and T. Hasama, 'Ultrafast all-optical refractive index modulation in intersub-band transition switch using InGaAs/AlAs/AlAsSb quantum well,' *Jpn. J. Appl. Phys.*, **46**, L157–L160 (2007).

[16] N. Suzuki and N. Iizuka, 'Effect of polarization field in intersub-band transition in AlGaN/GaN quantum wells,' *Jpn. J. Appl. Phys.*, **38**, L363–L365 (1999).

[17] Y. Hirayama, J. H. Smet, L.-H. Peng, C. G. Fonstad, and E. P. Ippen, 'Feasibility of 1.55 µm intersub-band photonic devices using InGaAs/AlAs pseudomorphic quantum well structures,' *Jpn J. Appl. Phys.*, **33**, 890–895 (1994).

[18] D. Ahn and S.-L. Chuang, 'Optical gain in a strained-layer quantum well laser,' *IEEE J. Quantum Electron.*, **24**, 2400–2406 (1988).

[19] B. F. Levine, 'Quantum-well infrared photodetectors,' *J. Appl. Phys.*, **74**, R1–R81 (1993).

[20] N. Suzuki, N. Iizuka, and K. Kaneko, 'FDTD simulation of femtosecond optical gating in nonlinear optical waveguide utilizing intersub-band transition in AlGaN/GaN quantum wells,' *IEICE Trans. Electron.*, **E83-C**, 981–988 (2000).

[21] R. Akimoto, B. S. Li, K. Akita, and T. Hasama, 'Ultrafast all-optical switching devices based on intersub-band transitions in II-VI quantum wells,' in *Proceedings the 30th European Conference on Optical Communication*, Vollume 4, Stockholm, Sweden, pp. 912–915 (2004).

[22] J. Hamazaki, H. Kunugita, K. Ema, A. Kikuchi, and K. Kishino, 'Intersub-band relaxation dynamics in GaN/AlN multiple quantum wells studied by two-color pump-probe experiments,' *Phys. Rev. B*, **71**, 165334 (2005).

[23] N. Suzuki and N. Iizuka, 'Feasibility study on ultrafast nonlinear optical properties of 1.55-µm intersub-band transition in AlGaN/GaN quantum wells,' *Jpn J. Appl. Phys.*, **36**, L1006–L1008 (1997).

[24] N. Suzuki and N. Iizuka, 'Electron scattering rates in AlGaN/GaN quantum wells for 1.55-µm inter-sub-band transition,' *Jpn J. Appl. Phys.*, **37**, L369–L371 (1998).

[25] N. Suzuki and N. Iizuka, 'AlGaN/GaN intersub-band transitoions for 1.55-µm optical switches,' in M. Osiński, P. Blood, and A. Ishibashi (Eds) *Physics and Simulation of Optoelectronic Devices VI*, San Jose, Proceedings SPIE, Volume 3283, Part 2, pp. 614–621 (1998).

[26] J. D. Heber, C. Gmachl, H. M. Ng, and A. Y. Cho, 'Comparative study of ultrafast intersub-band electron scattering times at ~1.55 µm wavelength in GaN/AlGaN heterostructures,' *Appl. Phys. Lett.*, **81**, 1237–1239 (2002).

[27] J. Hamazaki, S. Matsui, H. Kunugita, K. Ema, H. Kanazawa, T. Tachibana, A. Kikuchi, and K. Kishino, 'Ultrafast intersub-band relaxation and nonlinear susceptibility at 1.55 μm in GaN/AlN multiple-quantum wells,' *Appl. Phys. Lett.*, **84**, 1102–1104 (2004).

[28] N. Iizuka, K. Kaneko, and N. Suzuki, 'Sub-picosecond modulation by intersub-band transition in ridge waveguide with GaN/AlN quantum wells,' *Electron. Lett.*, **40**, 962–963 (2004).

[29] N. Iizuka, K. Kaneko, and N. Suzuki, 'Sub-picosecond all-optical gate utilizing GaN intersub-band transition ,' *Optics Express*, **13**, 3835–3840 (2005).

[30] L. Wendler and R. Pechstedt, 'Dynamical screening, collective excitations, and electron–phonon interaction in heterostructures and semiconductor quantum wells: Application to double heterostructure,' *Phys. Status Solidi (b)*, **141**, 129–150 (1987).

[31] K. Huang and B. Zhu, 'Dielectric continuum model and Fröhlich interaction in superlattices,' *Phys. Rev. B*, **38**, 13377–13386 (1988).

[32] M. Dür, S. M. Goodnick, and P. Lugli, 'Monte Carlo simulation of intersub-band relaxation in wide, uniformly doped GaAs/Al$_x$Ga$_{1-x}$ As quantum wells,' *Phys. Rev. B*, **54**, 17794–17804 (1996).

[33] N. Suzuki, N. Iizuka, and K. Kaneko, 'Simulation of ultrafast GaN/AlN intersub-band optical switches,' *IEICE Trans. Electron.*, **E88-C**, 342–348 (2005).

[34] N. Iizuka, K. Kaneko, and N. Suzuki, 'Polarization dependent loss in III-nitride optical waveguides for telecommunication devices,' *J. Appl. Phys.*, **99**, 093107 (2006).

[35] N. Suzuki, 'Intersub-band optical switches for optical communications,' in J. Piprek (Ed.) *Nitride Semiconductor Devices: Principles and Simulation*, Weinheim: Wiley–VCH, Chapter 11, pp. 235–252. (2007).

[36] F. Bernardini, V. Fiorentini, and D. Vanderbilt, 'Spontaneous polarization and piezoelectric constants of III-V nitrides,' *Phys. Rev. B*, **56**, R10024–R10027 (1997).

[37] N. Suzuki, N. Iizuka, and K. Kaneko, 'Calculation of near-infrared intersub-band absorption spectra in GaN/AlN quantum wells,' *Jpn J. Appl. Phys.*, **42**, 132–139 (2003).

[38] A. Harwit, J. S. Harris, Jr., and A. Kapitulnik, 'Calculated quasi-eigenstates and quasi-eigenenergies of quantum well superlattices in an applied electric field,' *J. Appl. Phys.*, **60**, 3211–3213 (1986).

[39] K. Kim, W. R. L. Lambrecht, and B. Segall, 'Elastic constants and related properties of tetrahedrally bonded BN, AlN, GaN, and InN,' *Phys. Rev. B*, **53**, 16310–16326 (1996).

[40] B. E. Foutz, L. F. Eastman, U. V. Bhapkar, and M. S. Shur, 'Comparison of high field electron transport in GaN and GaAs,' *Appl. Phys. Lett.*, **70**, 2849–2851 (1997).

[41] K. Hoshino, T. Someya, K. Hirakawa, and Y. Arakawa, 'Observation of intersub-band transition from the first to the third sub-band (e1-e3) in GaN/AlGaN quantum wells,' *Phys. Stat. Sol. (a)*, **192**, 27–32 (2002).

[42] I. Waki, C. Kumtornkittikul, Y. Shimogaki, and Y. Nakano, 'Shortest intersub-band transition wavelength (1.68 μm) achieved in AlN/GaN multiple quantum wells by metalorganic vapor phase epitaxy,' *Appl. Phys. Lett.*, **82**, 4465–4467 (2003); Erratum, *Appl. Phys. Lett.*, **84**, p. 3703 (2004).

[43] S. Nicolay, E. Feltin, J.-F. Carlin, M. Mosca, L. Nevou, M. Tchernycheva, F.H. Julien, M. Ilegems, and N. Grandjean, 'Indium surfactant effect on AlN/GaN heterostructures grown by metal-organic vapor-phase epitaxy: Applications to intersub-band transitions,' *Appl. Phys. Lett.*, **88**, 151902 (2006).

[44] C. Gmachl and H. M. Ng, 'Intersub-band absorption at λ ∼ 2.1 μm in A-plane GaN/AlN multiple quantum wells,' *Electron. Lett.*, **39**, 567–569 (2003).

[45] E. A. DeCuir, Jr., E. Fred, M. O. Manasreh, J. Schörmann, D. J. As, and K. Lischka, 'Near-infrared intersub-band absorption in nonpolar cubic GaN/AlN superlattices,' *Appl. Phys. Lett.*, **91**, 041911 (2007).

[46] M. Tchernycheva, L. Nevou, L. Doyennette, F. H. Julien, E. Warde, F. Guillot, E. Monroy, E. Bellet-Amalric, T. Remmele, and M. Albrecht, 'Systematic experimental and theoretical investigation of intersub-band absorption in GaN/AlN quantum wells,' *Phys. Rev. B*, **73**, 125347 (2006).

[47] E. A. DeCuir, Jr., E. Fred, B. S. Passmore, A. Muddasani, M. O. Manasreh, J. Xie, H. Mokoç, M. E. Ware, and G. J. Salamo, 'Near-infrared wavelength intersub-band transitions in GaN/AlN short period superlattices,' *Appl. Phys. Lett.*, **89**, 151112 (2006).

[48] M. Tchernycheva, L. Nevou, L. Doyennette, F. H. Julien, F. Guillot, E. Monroy, T. Remmele, and M. Albrecht, 'Electron confinement in strongly coupled GaN/AlN quantum wells,' *Appl. Phys. Lett.*, **88**, 153113 (2006).

[49] N. Iizuka, K. Kaneko, and N. Suzuki, 'All-optical switch utilizing intersub-band transition in GaN quantum wells,' *IEEE J. Quantum Electron.*, **42**, 765–771 (2006).

[50] A. Kikuchi, T. Yamada, S. Nakamura, S. Kusakabe, D. Sugihara, and K. Kishino, 'Improvement of crystal quality of RF-plasma-assisted molecular beam epitaxy grown Ga-polarity GaN by high-temperature grown AlN multiple intermediate layers,' *Jpn J. Appl. Phys.*, **39**, L330–L333 (2000).

[51] T. Shimizu, C. Kumtornkittikul, N. Iizuka, M. Sugiyama, and Y. Nakano, 'Intersub-band transition of AlN/GaN quantum wells in optimized AlN-based waveguide structure', *Conference on Lasers and Electro-Optics*, Baltimore (2007), JThD21.

[52] C. Kumtornkittikul, T. Shimizu, N. Iizuka, N. Suzuki, M. Sugiyama, and Y. Nakano, 'AlN waveguide with GaN/AlN quantum wells for all-optical switch utilizing intersub-band transition,' *Jpn J. Appl. Phys.*, **46**, L352–L355 (2007).

[53] Y. Li, A. Bhattacharyya, C. Thomidis, T. D. Moustakas, and R. Paiella, 'Nonlinear optical waveguides based on near-infrared intersub-band transitions in GaN/AlN quantum wells,' *Optics Express*, **15**, 5860–5865 (2007).

[54] P. Sanchis, J. Blasco, A, Martínez, and J. Martí, 'Design of silicon-based slot waveguide configurations for optimum nonlinear performance,' *J. Lightwave Technol.*, **25**, 1298–1305 (2007).

[55] N. Iizuka, K. Kaneko, and N. Suzuki, 'Ultrafast all-optical modulation by GaN intersub-band transition in a ridge waveguide,' in *2004 IEEE LEOS Annual Meeting, Conference Proceedings*, Volume 2, Puerto Rico (2004), pp. 665–666.

[56] N. Iizuka, K. Kaneko, and N. Suzuki, 'Time-resolved characterization of all-optical switch utilizing GaN intersub-band transition,' in *Abstract Int. Quantum Electronics Conference and the Pacific Rim Conference on Lasers and Electro-Optics*, Tokyo (2005), pp. 1279–1280.

[57] R. Akimoto, Y. Kinpara, K. Akita, F. Sasaki, and S. Kobayashi, 'Short-wavelength intersub-band transitions down to 1.6 μm in ZnSe/BeTe type-II superlattices', *Appl. Phys. Lett.*, **78**, 580–582 (2001).

[58] R. Akimoto, K. Akita, F. Sasaki, and S. Kobayashi, 'Short-wavelength ($\lambda < 2\,\mu$m) intersub-band absorption dynamics in ZnSe/BeTe quantum wells', *Appl. Phys. Lett.*, **80**, 2433–2435 (2002).

[59] C. Heske, U. Groh, O. Fuchs, L. Weinhardt, E. Umbach, M. Grün, S. Petillon, A. Dinger, C. Klingshirn, W. Szuszkiewicz, and A. Fleszar, 'Studying the local chemical environment of sulfur atoms at buried interfaces in CdS/ZnSe superlattices', *Appl. Phys. Lett.*, **83**,. 2360–2362 (2003).

[60] B. S. Li, R. Akimoto, K. Akita, and T. Hasama, 'ZnSe interlayer effects on properties of (CdS/ZnSe)/BeTe superlattices grown by molecular beam epitaxy', *J. Appl. Phys.*, **99**, 44912 (2006).

[61] M. Nagelstrasser, H. Droge, F. Fischer, T. Litz, A. Waag, G. Landwehr, H.-P. Steinruck, 'Band discontinuities and local interface composition in BeTe/ZnSe heterostructures', *J. Appl. Phys.*, **83**, 4253–4257 (1998).

[62] M. Goppert, M. Grun, C. Maier, S. Petillon, R. Becker, A. Dinger, A. Strozum, M. Jorger, and C. Klingshirn, 'Intersub-band and interminiband spectroscopy of doped and undoped CdS/ZnSe multiple quantum wells and superlattices', *Phys. Rev. B*, **65**, 115334 (2002).

[63] B. S. Li, R. Akimoto, K. Akita, and T. Hasama, 'Structural study of (CdS/ZnSe)/BeTe superlattices for $\lambda = 1.55\,\mu$m intersub-band transition', *J. Appl. Phys.*, **95**, 5352–5359 (2004).

[64] B. S. Li, R. Akimoto, K. Akita, and T. Hasama, 'Shorter wavelength intersub-band absorption down to $\lambda = 1.55\,\mu$m in (CdS/ZnSe)/BeTe type-II superlattices', *Phys. Stat. Sol.(c)*, **3**, 1147–1151 (2006).

[65] GW. Cong, R. Akimoto, K. Akita, T. Hasama, and H. Ishikawa, 'Low-saturation-energy-driven ultrafast all-optical switching operation in (CdS/ZnSe)/BeTe intersub-band transition', *Optics Express*, **15**, 12123–12130 (2007).

[66] K. Akita, R. Akimoto, T. Hasama, H. Ishikawa, and Y. Takanashi, 'Intersub-band all-optical switching in sub-micron high-mesa SCH waveguide structure with wide-gap II-VI-based quantum wells,' *Electron. Lett.*, **42**, 1352–1353 (2006).

[67] BeamProp™, Rsoft Design Group, Inc., 400 Executive Boulevard, Suite 100 Ossining, NY 10562.

[68] H. C. Liu and F. Capasso, *Intersub-band Transitions in Quantum Wells: Physics and Device Applications I*, Academic Press, New York, Chapter 1(2000).

[69] P. Harrison, *Quantum Wells, Wires and Dots: Theoretical and Computational Physics*, John Wiley & Sons Ltd, Chichester, Chapter 3 (2000).

[70] R. Akimoto, B.S. Li, K. Akita, and T. Hasama, 'Subpicosecond saturation of intersub-band absorption in (CdS/ZnSe)/BeTe quantum-well waveguides at telecommunication wavelength,' *Appl. Phys. Lett.*, **87**, 181104 (2005).

[71] H. Ishikawa, R. Akimoto, Ultrafast all-optical switches using intersub-band transition in quantum well,' *Oyo Butsuri*,**76** (3), 291–295 (2007) (in Japanese).

[72] N. Georgiev, T. Mozume, 'Photoluminescence study of InGaAs/AlAsSb heterostructures,' *J. Appl. Phys.*, **89**, 1064–1069 (2001).

[73] T. Mozume, N. Georgiev, 'Optical and structural characterization of InGaAs/AlAsSb quantum wells grown by molecular beam epitaxy,' *Jpn J. Appl. Phys.*, **41**, Part1, No.2B, 1008–1011 (2002).

[74] H. Tsuchida, T. Simoyama, H. Ishikawa, T. Mozume, M. Nagase, and J. Kasai, 'Cross-phase-modulation based wavelength conversion using intersub-band transition in InGaAs/AlAs/AlAsSb coupled quantum wells.' *Opt. Lett.*, **32**, 751–753 (2007).

[75] E. O. Kane, 'Band structure of indium antimonide', *J. Phys. Chem. Solids*, **1**, 249–261 (1957).

[76] T. Asano, S. Noda, T. Abe, and A. Sasaki, 'Investigation of short wavelength intersub-band transitions in INGaAs/AlAs quantum wells on GaAs substrate,' *J. Appl. Phys.*, **82**, 3385–3391, (1997).

[77] M. Sugawara, N. Okazaki, T. Fujii, and S. Yamazaki, 'Conduction-band and valence-band structures in strained $In_{1-x}Ga_xAs/InP$ quantum wells on (0 0 1) InP substrates', *Phys. Rev. B*, **48**, 8102–8118 (1993).

[78] H. Haug and S. W. Koch, '*Quantum Theory of the Optical and Electrical Properties of Semiconductors*', World Scientific Publishing Co. Pte Ltd, Singapore, Third edition, p.149 (1994).

[79] R. Akimoto, T. Simoyama, H. Tsuchida, S. Namiki, C. G. Lim, M. Nagase, T. Mozume, T. Hasama, and H. Ishikawa, 'All-optical demultiplexing of 160–10 Gbit/s signals with Mach–Zehnder interferometroic switch utilizing intersub-band transition in InGaAs/AlAs/AlAsSb quantum well,' *Appl. Phys. Lett.*, **91**, 221115 (2007).

6

Wavelength Conversion Devices

Haruhiko Kuwatsuka

6.1 Introduction

Wavelength conversion is one of the most important functions of optical signal processing in future photonic network systems [1, 2]. It is indispensable for wavelength routing in WDM (Wavelength Division Multiplexing) systems. Wavelength conversion is also important in ultra-fast OTDM (Optical Time Division Multiplexing) systems [3]. For example, it is needed to adapt the wavelength of optical signals to a wavelength providing suitable chromatic dispersion of optical fibers, because the chromatic dispersion seriously influences the transmission characteristics in ultra-fast OTDM systems. In a combination of OTDM and WDM, wavelength conversion is also essential.

All-optical wavelength conversion is also required to overcome the recent severe power consumption problems of communication equipment. All-optical wavelength conversion allows us to realize optical systems without electronic circuits, which consume large amounts of electric power. In addition, semiconductor devices for wavelength conversion are desirable because of their potential high speed, compactness, low-power consumption, and the possibility of integration with other semiconductor active devices, such as semiconductor laser diodes, photo-detectors, and so on.

Various methods of all-optical wavelength conversion have been proposed to replace the method of modulation of different wavelength light sources after an OE (Opto-Electro) conversion. In this chapter, we at first survey various types of all-optical wavelength conversion using semiconductor optical devices and other materials. We then focus on wavelength conversion using FWM (Four-Wave Mixing) in semiconductor optical devices. The principle of FWM in semiconductor laser diodes (LDs) or SOAs (Semiconductor Optical Amplifiers) is discussed. The physics of wavelength conversion of short pulses is also described. Several experiments

Ultrafast All-Optical Signal Processing Devices Edited by Hiroshi Ishikawa
© 2008 John Wiley & Sons, Ltd

and applications of wavelength conversion using FWM in LDs or SOAs are presented. The future of wavelength conversion using FWM is described at the end.

6.2 Wavelength Conversion Schemes

There are various methods of wavelength conversion in addition to that of modulating different wavelength light sources after OE conversion. These include modulating different wavelength light sources by an optical gate switch, and coherent wavelength conversion generating a new light with a different wavelength using second-order nonlinear effects or third-order nonlinear effects. Table 6.1 shows a short categorization of wavelength conversion using semiconductor devices and other materials.

6.2.1 Optical Gate Switch Type

Optical gate, switch-type wavelength converters are based on optical nonlinearity of mainly optical fibers or semiconductors. The merit of optical fiber is its fast response time of less than a picosecond due to its optical Kerr effect. Although its third-order nonlinear susceptibility is small, we can realize a sufficient level of speed and extinction ratio because we can use a very long interaction length, for example, tens of meters owing to its low dispersion and low loss characteristics. Proposals are the use of Kerr rotation [4] or XPM (Cross Phase Modulation) of NOLM (Nonlinear Optical Loop Mirror) [5], etc. For this purpose, highly nonlinear optical fiber has been developed, which has a one order of magnitude larger optical confinement to the core than that of normal optical fiber [6].

In a semiconductor device, there are various proposals for optical gate-switch type wavelength converters. In the case of SOA, two types of mechanisms are used. One is XPM and the other is XGM (cross gain modulation). Interferometric type devices like a MZI (Mach–Zehnder interferometer) [7], described in Chapter 3, or a DISC (Delayed Interference Signal Converter) [8], can be used to utilize the XPM effect in SOAs. Other than SOAs, various kinds of optical gate-switch can be used such as that using absorption saturation of ISBT (Inter Sub-Band Transition) [9], which is described in Chapter 5, and the one using two-photon absorption and so on.

It has been verified that a MZI can convert the wavelength of signals of over a bit-rate of 160 Gbps by a push/pull action of inputting control light with a time difference into both the arms of an SMZ (Symmetric Mach–Zehnder interferometer) [10]. Since input/output characteristics of an SMZ are sinusoidal and nonlinear, there is waveform-shaping function in an SMZ wavelength converter. However, this at the same time makes the operating conditions of the interferometer critical. Precise adjustment of the phase shift of the interferometer is required.

A DISC is a kind of differential mode interferometic switch for canceling the slow recovery of the carrier density in an SOA. After a CW probe is modulated by an input signal in an SOA, the slow component of the modulated probe is compensated by an AMZI (Asymmetric Mach–Zehnder interferometer) filter in which the differential delay time of the two arms is controlled in order to flatten the small signal frequency response of XPM in the SOA. Operation at bit-rates beyond 100 Gbps has been confirmed.

Table 6.1 Comparison of various wavelength conversion methods

	Principle	Device	Speed	Modulation method	Input/output ratio	Extinction ratio	Linearity	Input level	Wavelength dependence	Output wavelength variability	Integration	Remarks
OE-EO type	Opto-electric conversion	PD + Electric + LD	40 Gb/s	Format fixed	Per electric amplifier	OK	OK	OK	Per light source	OK	Difficult	
Optical gate switch type	Cross phase modulation (XPM)	Optical fiber	~1 Tb/s	Format fixed	8 dB	>40 dB	NG	+10 dBm	None	OK	Difficult	Waveform shaping Optical filter needed
		SOA-MZI or DISC	~160 Gb/s	Format fixed	10 dB	20 dB Dependent on operation point	NG	-15 ~ -5 dBm	Weak	OK	Integration device	Waveform shaping Negative chirp operation Optical fiber unneeded
	Cross gain modulation (XGM)	SOA	~20 Gb/s	OOK	3 dB	18 dB	NG	-10 ~ 0 dBm	Medium	OK	Easy	ON/OFF Reversal Optical filter unneeded
Coherent type	Difference Frequency Generation (DFG)	PPLN	>1 Tb/s	Format non-dependent	-20 dB	>40 dB	OK	-30 ~ 10 dBm	None	Two times tunable wavelength laser	Difficult	Optical filter needed Conjugate light generation
	Non-degenerate four-wave mixing (ND-FWM)	Optical fiber	>1 Tb/s	Format non-dependent	-10 dB	>40 dB	OK	-20 ~ 0 dBm	None	Two times tunable wavelength laser	Difficult	Optical filter needed Conjugate light generation
		SOA or LD	~1 THz	Format non-dependent	+5 dB	+30 dB	OK	-25 ~ 0 dBm	Strong	Two times tunable wavelength laser	Easy	Optical filter needed Conjugate light generation

The XGM in SOA is the effect where optical gain at a wavelength decreases when the optical gain is saturated by the optical signal with another wavelength. Although the configuration of a wavelength converter using XGM in SOA is simple, the logic of the converted signal is inversed and the speed of wavelength conversion is limited by the carrier lifetime in SOAs, which is several nanoseconds. Only OOK (on/off keying) format signals can be converted. To return the inverted logic to the original, cascaded configuration of SOAs has been proposed. The merit of simple configuration of XGM wavelength conversion is, however, decreased by the cascaded configuration.

Generally, in optical gate-switch type wavelength conversion, the intensity of output light is not linear to that of the input signal light. As a result, the input power range is narrow in optical gate-switch type wavelength conversion. On the other hand, this gives a reshaping effect to the wavelength-converted optical pulse. One advantage of the optical gate type switch is that there can be a structure requiring no wavelength filter to separate the input signal light and the converted output light, unlike the coherent-type wavelength converter to be discussed later.

6.2.2 Coherent Type Conversion

The coherent type, unlike the optical gate switch type, can be used not only for conventional NRZ (non-return to zero) and OOK format but also for various formats such as PSK (phase shift keying), FSK (frequency shift keying), multi-level modulation formats including duo-binary modulation format, QPSK, QAM and so on [11]. This format-free characteristic of the coherent types is of great importance in future network systems, where various modulation formats are to be used. In addition, the linearity of converted output light to the input signal light is better than that of optical gate switch type. Wavelength conversion on a bundle of WDM signals is also possible and has been demonstrated [6]. Its shortcoming is the need for an optical filter to separate the output light and input signal light at the end. To convert the wavelength to an arbitrary wavelength, it is necessary to change not only the pump wavelength, but also the wavelength of the filter.

As a device for difference frequency generation, an optical waveguide made of LiNbO$_3$ with periodically poled domain (PPLN) is used. The periodically poled domain is for phase matching between the pump, signal, and converted wave [12]. As a pump, 0.78 μm wavelength light is used. In some cases, the pump wave is supplied through virtual generation of a wave by injecting light of 1.55-μm wavelength, not light of 0.78-μm wavelength. Because of the phase matching condition, the selection of the pump wavelength is critical.

As for four-wave mixing (FWM), there are several reports [6] of experiments with optical fibers using their low dispersion and low loss characteristics. The fiber devices have exhibited sufficient characteristics. For semiconductor devices, there is a number of reports on four-wave mixing by using an active layer of a SOA [13–27] or a LD [28–35]. Conversion efficiency greater than unity can be obtained using a short semiconductor device of several hundred μm in length, owing to large optical-nonlinear susceptibility and large optical gain. The optical gain solves the problem of trade-off between the larger absorption and larger optical-nonlinear susceptibility in the usual absorptive optical-nonlinear materials. As interaction length is shorter than several mm, the consideration of phase match conditions is not required, in contrast to the case of FWM in optical fibers. The wavelength of the pump and signal is selected optionally. Monolithic integration with a pump light source is also feasible. Since the degradation of NF (noise figure) arises from ASE (amplified spontaneous emission), it is necessary to achieve large conversion efficiency with small optical gain. In addition, a realization of higher efficiency conversion over the whole C-band is also required.

In order to realize small wavelength conversion systems for future wavelength cross-connection of WDM signals, arrayed integration of wavelength converters with a wavelength-tunable light source will be necessary. If the performance of semiconductor-based coherent type devices reaches a practical level and integration with a wavelength-tunable light source is realized, it will have a huge impact on network architectures in the future.

6.3 Physics of Four-wave Mixing in LDs or SOAs

As described in the previous section, FWM in LDs or SOAs is a promising method for achieving high speed, wide range, and transparent wavelength conversion. Wavelength conversion of ultra-short pulses of several ps and high-speed signals with bit rates above 160 Gbps are demonstrated. Also demonstrated are format-free wavelength conversions of OOK and DPSK signals. In addition, FWM can be used for phase-conjugate wave generation. Phase conjugate optical light generation by FWM can compensate even-order dispersions among high-order dispersions of the optical fiber. Semiconductors under population inversion conditions, which are used for active regions in these devices, are suitable media for FWM in 1.3–1.6 μm wavelength range for optical communication. They have large third order nonlinear optical susceptibilities $\chi^{(3)}$ of more than 1×10^{-15} m^2/V^2 [36], which are obtained by resonating the band gap energy with the photon energy of pump, signal and conjugate beams. The intensity of the conjugate beams obtained with a short interaction length of less than 1 mm is sufficient for practical use. As carrier-injected semiconductors have large linear optical gain, we can avoid the severe optical loss that usually occurs under resonant conditions in the absorptive medium. The conjugate beam is also amplified by the linear gain.

In this section, we discuss the mechanism of FWM in semiconductors under population inversion conditions using a semi-classical model. The model gives appropriate $\chi^{(3)}$ values enough for explaining the conversion efficiency using FWM in SOA or LDs. The model also explains the asymmetric $\chi^{(3)}$ for positive and negative detuning in SOAs with a MQW or bulk active layer.

6.3.1 Model

Figure 6.1 illustrates the principle of wavelength conversion using the FWM. When two strong pump waves (frequencies f_{p1} and f_{p2}) are put into a SOA or a LD with an input signal (frequency f_s), the conjugate wave (frequency $f_{c2} = f_{p1} + f_{p2} - f_s$) of the signal is generated by the third order nonlinear effect. In addition, the replica waves (frequency $f_r = f_{p1} - f_{p2} + f_s$, or $f_r = f_{p2} - f_{p1} + f_s$) are also generated. We can convert the wavelength of the signal ($\lambda_p = c/f_p$) to the wavelength of the conjugate ($\lambda_{c2} = c/f_{c2}$) or to the replica ($\lambda_r = c/f_r$). In the special case of pump 1 and pump 2 being degenerate, where only one pump wave is used, the wavelength can be converted to the wavelength of conjugate ($\lambda_{c1} = c/f_{c1}$) next to the pump.

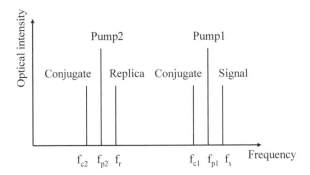

Figure 6.1   The principle of wavelength conversion by four-wave mixing

Three kinds of third order nonlinear optical effects contributing to FWM occur in semiconductor gain media [34]. They are the carrier-density pulsation (CDP) effect, the carrier heating (CH) effect, and the spectral-hole burning (SHB) effect. The discussion using density-matrix formalism can explain the FWM in semiconductor gain media including the undesired asymmetric nonlinear gain and asymmetric conversion efficiency in the pump-signal detuning dependence [36].

First, we discuss the CDP effect. Beating between the pump and signal causes pulsation of carrier density and the pulsation of Fermi energy as shown in Figure 6.2(a). This results in the pulsation of the gain and refractive index. The effect can be explained by conventional rate equation [36] and described as:

$$\chi_{CDP}^{(3)}\left(\omega_c; \omega_{p1}, \omega_{p2}, -\omega_s\right) = (i - \alpha_c)\left(\frac{dg}{dN}\right) g \frac{\varepsilon\varepsilon_0 \hbar c_0^2}{3\left(\hbar\omega_i\right)\left(\hbar\omega_p\right)} \frac{1}{i\left(\omega_{p1} - \omega_s\right) + \left(\frac{dg}{dN}\right)\frac{\eta\varepsilon\varepsilon_0 c_0}{2\hbar\omega_p}E^2 + \frac{1}{\tau_c}}$$

$$(6.1)$$

for the conjugate generation. In the degenerate case of pump 1 and pump 2, i.e., ω_{p1} is equal to ω_{p2}. We obtain $\chi^{(3)}$ for the generation of replica frequency ω_r, which is not a conjugate wave but has the same phase as the wave of ω_s.

$$\chi_{CDP}^{(3)}\left(\omega_r; -\omega_{p1}, \omega_{p2}, \omega_s\right) = (i - \alpha_c)\left(\frac{dg}{dN}\right) g \frac{\varepsilon\varepsilon_0 \hbar c_0^2}{3\left(\hbar\omega_i\right)\left(\hbar\omega_p\right)} \times$$

$$\frac{1}{-i\left(\omega_{p1} - \omega_s\right) + \left(\frac{dg}{dN}\right)\frac{\eta\varepsilon\varepsilon_0 c_0}{2\hbar\omega_p}E^2 + \frac{1}{\tau_c}}$$

$$(6.2)$$

where g is the linear gain, N the carrier density, ε the dielectric constant of the material, ε_0 the permittivity in vacuum, $\hbar$ the reduced Planck constant, c_0 the speed of light in vacuum, ω_c, ω_r, ω_{p1}, and ω_{p2} the angular frequency of the conjugate, the replica, pump 1 and pump 2, respectively, $\Delta\omega$, the detuning angular frequency between the pump and the signal, η, the refractive index, E the optical electric field, and τ_c is the carrier lifetime. α_c is the line-width enhancement factor due to the carrier density change, described as:

$$\alpha_c = -2\frac{\omega}{c_0}\left(\frac{d\eta}{dN}\right)\bigg/\left(\frac{dg}{dN}\right) \tag{6.3}$$

As α_c is bigger than unity in bulk or quantum well (QW) semiconductors, the real number part of $\chi_{CDP}^{(3)}$ is larger than the imaginary number part. The CDP effect corresponds to the band filling effect in the nonpopulation-inversion case. The bandwidth is not limited by the spontaneous emission carrier-life-time. It is determined by the stimulated emission due to the strong pump field. The bandwidth is an order of several 10 GHz. In the case of FWM, the detuning exceeds several 100 GHz and $\chi_{CDP}^{(3)}$ decreases proportionally to the inverse of the detuning.

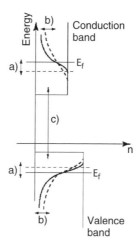

Figure 6.2 Three third-order nonlinear effects in a semiconductor active region in SOAs or LDs. (a) Carrier density pulsation (CDP) effect; (b) carrier heating (CH) effect, and (c) spectral hole burning (SHB) effect [36]. (Reproduced by permission of © 1999 IEEE)

Next, we discuss the CH effect. As the energy of the carriers generated by the beating between the pump and the signal is different from the quasi-Fermi energy, the beating modulates the temperature of the carriers as shown by (b) in Figure 6.2. The modulation of temperature results in the pulsation of the refractive index and the linear gain. From classical analysis [36], the CH component of $\chi^{(3)}$ for conjugate generation is described by:

$$\chi_{CH}^{(3)}\left(\omega_c;\omega_{p1},\omega_{p2},-\omega_s\right) = (\alpha_h - i)g\frac{c_0^2\hbar\varepsilon\varepsilon_0\Delta E}{3(\hbar\omega)^2}\left(\frac{dg}{dU}\right)_{N=N_0}\frac{\tau_h}{i\left(\omega_{p1}-\omega_s\right)\tau_h+1} \tag{6.4}$$

in the degenerate case of pump 1 and pump 2, i.e., ω_{p1} equal to ω_{p2}. The $\chi^{(3)}$ for the replica generation is:

$$\chi_{CH}^{(3)}\left(\omega_r; -\omega_{p1}, \omega_{p2}, \omega_s\right) = (\alpha_h - i)g\frac{c_0^2\hbar\varepsilon\varepsilon_0\Delta E}{3(\hbar\omega)^2}\left(\frac{dg}{dU}\right)_{N=N_0}\frac{\tau_h}{-i\left(\omega_{p1} - \omega_s\right)\tau_h + 1} \quad (6.5)$$

where U is the total energy of carriers defined as:

$$U = \int d\mathbf{k}\, E_c f_c(E_c(\mathbf{k})) + E_v f_v(E_v(\mathbf{k})) \quad (6.6)$$

ΔE is the energy change caused by the generation of a photon, which is given by:

$$\Delta E = \hbar\omega - (F_c + F_v + E_g) - \left(\frac{dU}{dN}\right)_{T=T_0} \quad (6.7)$$

where F_c and F_v are quasi-Fermi energies of electrons and holes. τ_h is the relaxation time of the carrier distribution heated by the CH effect. It is usually determined by LO (longitudinal optical) phonon scattering time of carriers. α_h is the line-width enhancement factor due to the CH effect, defined as:

$$\alpha_h = -2\frac{\omega}{c_0}\left(\frac{d\eta}{dT}\right)_{N=N_0}\Bigg/\left(\frac{dg}{dT}\right)_{N=N_0} \quad (6.8)$$

The CH effect is similar to the CDP effect except for the bandwidth that is determined by the LO-phonon scattering time. The LO-phonon scattering time in InGaAsP is several hundred femtoseconds. Then, the bandwidth of the CH component of the $\chi^{(3)}$ is several hundred GHz. In the bandwidth, the real number part of $\chi^{(3)}$ by the CH effect is larger than the imaginary number part. Above the bandwidth, $\chi^{(3)}$ by the CH effect decreases inversely in proportion to the detuning between the pump and the signal. The imaginary number part is larger than the real number part in this region. The sign of the imaginary number part is opposite in positive and negative detuning.

The third is the SHB effect as shown by (c) in Figure 6.2. The SHB effect is the result of the summation of the individual nonlinear response of the dipole in the band. Using the density matrix formalism under the relaxation time constant approximation [36], the third optical nonlinear susceptibility is obtained as:

$$\chi^{(3)}_{SHB}(\omega_c; \omega_{p1}, \omega_{p2}, -\omega_s)$$

$$= \sum_{j=1,2} \frac{e^4}{\varepsilon_0} \int d\mathbf{k} \frac{|<\Psi_c|r|\Psi_v>|^4 (f_v(\mathbf{k}) - f_c(\mathbf{k}))}{(\hbar\omega_c - E_c(\mathbf{k}) + E_v(\mathbf{k}) + i\hbar\Gamma_{cv})(\hbar\omega_{pj} - \hbar\omega_s + i\hbar\Gamma_{cc})(\hbar\omega_s - E_c(\mathbf{k}) + E_v(\mathbf{k}) - i\hbar\Gamma_{cv})}$$

$$+ \sum_{j=1,2} \frac{e^4}{\varepsilon_0} \int d\mathbf{k} \frac{|<\Psi_c|r|\Psi_v>|^4 (f_v(\mathbf{k}) - f_c(\mathbf{k}))}{(\hbar\omega_c - E_c(\mathbf{k}) + E_v(\mathbf{k}) + i\hbar\Gamma_{cv})(\hbar\omega_{pj} - \hbar\omega_s + i\hbar\Gamma_{cc})(\hbar\omega_{pj} - E_c(\mathbf{k}) + E_v(\mathbf{k}) - i\hbar\Gamma_{cv})}$$

$$+ \sum_{j=1,2} \frac{e^4}{\varepsilon_0} \int d\mathbf{k} \frac{|<\Psi_c|r|\Psi_v>|^4 (f_v(\mathbf{k}) - f_c(\mathbf{k}))}{(\hbar\omega_c - E_c(\mathbf{k}) + E_v(\mathbf{k}) + i\hbar\Gamma_{cv})(\hbar\omega_{pj} - \hbar\omega_s + i\hbar\Gamma_{vv})(\hbar\omega_s - E_c(\mathbf{k}) + E_v(\mathbf{k}) - i\hbar\Gamma_{cv})}$$

$$+ \sum_{j=1,2} \frac{e^4}{\varepsilon_0} \int d\mathbf{k} \frac{|<\Psi_c|r|\Psi_v>|^4 (f_v(\mathbf{k}) - f_c(\mathbf{k}))}{(\hbar\omega_c - E_c(\mathbf{k}) + E_v(\mathbf{k}) + i\hbar\Gamma_{cv})(\hbar\omega_{pj} - \hbar\omega_s + i\hbar\Gamma_{vv})(\hbar\omega_{pj} - E_c(\mathbf{k}) + E_v(\mathbf{k}) - i\hbar\Gamma_{cv})}$$

$$(6.9)$$

for conjugate generation. In the degenerate case, ω_{p1} is equal to ω_{p2}. The $\chi^{(3)}$ for the replica generation is given as:

$$\chi^{(3)}_{SHB}(\omega_r; \omega_{p1}, \omega_{p2}, -\omega_s)$$

$$= \sum_{j=1,2} \frac{e^4}{\varepsilon_0} \int d\mathbf{k} \frac{|<\Psi_c|r|\Psi_v>|^4 (f_v(\mathbf{k}) - f_c(\mathbf{k}))}{(\hbar\omega_c - E_c(\mathbf{k}) + E_v(\mathbf{k}) + i\hbar\Gamma_{cv})(-\hbar\omega_{pj} + \hbar\omega_s + i\hbar\Gamma_{cc})(\hbar\omega_s - E_c(\mathbf{k}) + E_v(\mathbf{k}) - i\hbar\Gamma_{cv})}$$

$$+ \sum_{j=1,2} \frac{e^4}{\varepsilon_0} \int d\mathbf{k} \frac{|<\Psi_c|r|\Psi_v>|^4 (f_v(\mathbf{k}) - f_c(\mathbf{k}))}{(\hbar\omega_c - E_c(\mathbf{k}) + E_v(\mathbf{k}) + i\hbar\Gamma_{cv})(-\hbar\omega_{pj} + \hbar\omega_s + i\hbar\Gamma_{cc})(\hbar\omega_{pj} - E_c(\mathbf{k}) + E_v(\mathbf{k}) - i\hbar\Gamma_{cv})}$$

$$+ \sum_{j=1,2} \frac{e^4}{\varepsilon_0} \int d\mathbf{k} \frac{|<\Psi_c|r|\Psi_v>|^4 (f_v(\mathbf{k}) - f_c(\mathbf{k}))}{(\hbar\omega_c - E_c(\mathbf{k}) + E_v(\mathbf{k}) + i\hbar\Gamma_{cv})(-\hbar\omega_{pj} + \hbar\omega_s + i\hbar\Gamma_{vv})(\hbar\omega_s - E_c(\mathbf{k}) + E_v(\mathbf{k}) - i\hbar\Gamma_{cv})}$$

$$+ \sum_{j=1,2} \frac{e^4}{\varepsilon_0} \int d\mathbf{k} \frac{|<\Psi_c|r|\Psi_v>|^4 (f_v(\mathbf{k}) - f_c(\mathbf{k}))}{(\hbar\omega_c - E_c(\mathbf{k}) + E_v(\mathbf{k}) + i\hbar\Gamma_{cv})(-\hbar\omega_{pj} + \hbar\omega_s + i\hbar\Gamma_{vv})(\hbar\omega_{pj} - E_c(\mathbf{k}) + E_v(\mathbf{k}) - i\hbar\Gamma_{cv})}$$

$$(6.10)$$

where only resonant terms are considered. e is the elementary charge, $\mathbf{k}$ is the wave vector of electrons and holes, Ψ_c and Ψ_v are the wave functions of the conduction band and the valence band, $f_v(\mathbf{k})$ and $f_c(\mathbf{k})$ are the Fermi distribution functions of electrons and holes, $E_c(\mathbf{k})$ and $E_v(\mathbf{k})$ are the energies of electrons and holes. By a detailed calculation, it can be understood that the imaginary number part of $\chi^{(3)}_{SHB}$ is larger than the real number part. As its sign is positive when the energy of all optical beams is in the bandwidth, this effect decreases the gain. The relaxation rate of the dipole Γ_{cv} and the dephasing rate of the conduction and valence band states Γ_{cc} and Γ_{vv} are determined by a fast process such as carrier–carrier scattering.

The carrier–carrier scattering rate determining Γ_{cv}, Γ_{cc} and Γ_{vv} is decreased when the density of states decreases. We can calculate the carrier–carrier scattering rate in electron-hole plasma in the calculated band structure. By treating the screening of the Coulomb and Fröhlich interactions in the fully dynamic RPA (random-phase approximation) [37], the scattering rate is given by:

$$\frac{1}{\tau(\mathbf{k})} = \frac{1}{\tau_{out}(\mathbf{k})} + \frac{1}{\tau_{in}(\mathbf{k})} \tag{6.11}$$

$$\frac{1}{\tau_{out}(\mathbf{k})} = \frac{(2\pi)}{\hbar} \frac{2}{(2\pi)^2} \int d\mathbf{q}[1 - f(\mathbf{k}-\mathbf{q})][n_b(\omega)+1]\frac{e^2}{|\mathbf{q}|} F(\mathbf{q})\frac{1}{2\pi}\mathrm{Im}\left(\frac{-1}{\varepsilon(\mathbf{q},\omega)}\right) \tag{6.12}$$

$$\frac{1}{\tau_{in}(\mathbf{k})} = \frac{(2\pi)}{\hbar} \frac{2}{(2\pi)^2} \int d\mathbf{q}[f(\mathbf{k}-\mathbf{q})][n_b(\omega)]\frac{e^2}{|\mathbf{q}|} F(\mathbf{q})\frac{1}{2\pi}\mathrm{Im}\left(\frac{-1}{\varepsilon(\mathbf{q},\omega)}\right) \tag{6.13}$$

where $\mathbf{q}$ is the difference of the wave vector between initial and final states, $\hbar\omega$ is the energy difference between initial and final states, and $n_b(\omega)$ is the Bose distribution function. The function $F(\mathbf{q})$ represents a form factor involving the overlap of the wave functions $\phi(z)$:

$$F(\mathbf{q}) = \iint dz_1 dz_2 \phi^*(z_1) \phi(z_1) \exp(q|z_1 - z_2|)\phi^*(z_2) \phi(z_2) \tag{6.14}$$

The function $\varepsilon(\mathbf{q}, \omega)$ is the complex RPA dielectric function, described as:

$$\varepsilon(\mathbf{q}, \omega) = \varepsilon_0 \varepsilon_\infty + \varepsilon_0 \frac{(\varepsilon_\infty - \varepsilon_s)\omega_{TO}^2}{\omega^2 - \omega_{TO}^2} - \varepsilon_p(\mathbf{q}, \omega) \tag{6.15}$$

where ε_∞ is the optical dielectric constant, ε_s is the static dielectric constant, and ω_{TO} is the TO (transverse optical) phonon angular frequency. The dynamic screening due to the carrier $\varepsilon p(\mathbf{q}, \omega)$ is given by:

$$\varepsilon_p(\mathbf{q}, \omega) = -\frac{2e^2}{|\mathbf{q}|} F(\mathbf{q}) \sum_\sigma \int d\mathbf{k} \frac{f_\sigma(\mathbf{k}+\mathbf{q}) - f_\sigma(\mathbf{k})}{E^\sigma(\mathbf{k}+\mathbf{q}) - E^\sigma(\mathbf{k}) - \hbar\omega - i\alpha} \tag{6.16}$$

where σ electrons or holes, $E(\mathbf{k})$ is the energy of carriers. The scattering rate given by Equations (6.1)–(6.16) includes electron–electron scattering (hole–hole scattering in the case of valence band), electron–hole scattering, and LO phonon scattering.

The calculated values for the scattering rate of electrons and holes using Equations (6.11)–(6.16) are 0.2×10^{14} to 1.0×10^{14} s^{-1} [36], which give enough response speed for the SHB effect for sub-picosecond wavelength conversion and sufficient bandwidth to cover the C-band in WDM systems.

6.3.2 Asymmetric $\chi^{(3)}$ for Positive and Negative Detuning

The interference effect between the CDP + CH effect and the SHB effect causes asymmetry for positive and negative detuning as shown in Figure 6.3. In the bandwidth of the SHB effect of several 10 THz, in which the detuning between pump and signal is less than 30 nm, the imaginary number part is dominant as expected from Equations (6.9) and (6.10). The sign is positive as shown by the gray arrows in Figure 6.3 for both cases of negative detuning (a) and positive detuning (b). For the CDP effect, the real number part is dominant when the detuning is less than the bandwidth determined by the carrier lifetime τ_c, as expected from Equations (6.1) and (6.2) because the line-width enhancement factor α_c is larger than unity. Total $\chi^{(3)}$ is the same for negative detuning (a) and positive detuning (b) as shown by broken white arrows. When the detuning is larger than the bandwidth of the CDP effect, which is the usual case for wavelength conversion of more than 1 nm, the imaginary number part is dominant for the CDP effect. The sign is negative for negative detuning as shown by the sold white arrow in Figure 6.3 (a) and positive for positive detuning as shown in (b). When the detuning is negative, the CDP effect and the SHB effect suppress each other, and the total of $\chi^{(3)}$ becomes small as shown by the black arrow in Figure 6.3 (a). When the detuning is positive as shown in Figure 6.3(b), the CDP + CH effect and the SHB effect strengthen each other. The total of $\chi^{(3)}$ becomes large as shown by the black arrow in Figure 6.3(b). This interference between the CDP effect and the SHB effect gives the asymmetric gain affecting the lasing characteristics of LDs and the asymmetric conversion efficiency for FWM wavelength conversion.

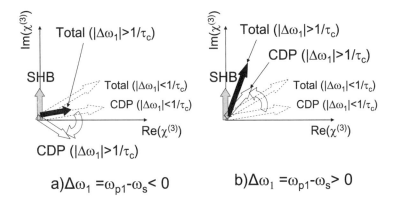

Figure 6.3   The interference between CDP+CH effect and the SHB effect. (a) The detuning is negative. (b) The detuning is positive

The example of calculated $\chi^{(3)}$ due to the CDP effect, the CH effect, the SHB effect, using Equations (6.1)–(6.16) are shown in Figure 6.4. For χ_h in the CH effect, an LO scattering time of 100 fs was used. The carrier density is 2.4×10^{18} cm^{-3}. In the detuning of $\Delta\omega < 1.0 \times 10^{12}$ rad/s, the absolute values of the $\chi^{(3)}$ due to the SHB effect are 5.6×10^{-16} m^2/V^2. The real number part of $\chi_{SHB}^{(3)}$ is negligible and the imaginary number part is dominant in this small detuning range. The imaginary number part corresponds to the gain saturation effect. From these $\chi_{SHB}^{(3)}$ values, the gain saturation coefficients are estimated to be 3.0×10^{-17} cm^3. In the degenerate case of pump and signal, the real number part corresponds to the Kerr effect. The real number part of the $\chi^{(3)}$ is -1.6×10^{-15} m^2/V^2, which corresponds to n_2 values of -5.8×10^{-11} cm^2/W. As the signs are negative, the CH effect corresponds to the self-defocusing effect. The real number part of $\chi_{SHB}^{(3)}$ is negligible in the degenerate case. The refractive index change due to the CDP effect in the degenerate case of pump and signal is not proportional to the optical power because the carrier recombination rate is changed by the optical power change. The Kerr effect due to the CDP effect cannot be defined.

From the measured conversion efficiency in four-wave mixing in a DFB (distributed feedback) LD, $\chi^{(3)}$ values for InP/InGaAs MQWs have been estimated [38]. Figure 6.5 shows the results. The asymmetry between the positive and negative detuning exists as expected from

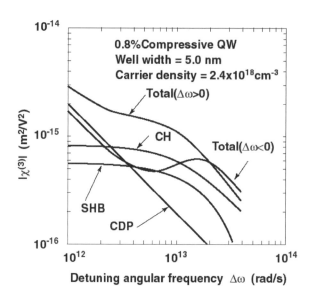

Figure 6.4 An example of the calculated $\chi^{(3)}$ for the detuning between the pump and the signal [36]. The degenerate case of pump 1 and pump 2 is shown [36]. (Reproduced by permission of © 1999 IEEE)

the discussion of the interference between the CDP + CH effect and the SHB effect. When the CDP effect and CH effect are described by the function of $a/(i\Delta\omega)$ and the SHB effect is described by the function of $(b + ci)/(1 + i\Delta\omega\tau)$, where b represents a phase effect equal to the sum of the carrier density modulation effect and carrier heating effect, and c represents the $\pi/2$ phase shift effect, $\chi^{(3)}$ can be described by the sum of both effects. It is described as:

$$\left|\chi^{(3)}\right| = \left|\frac{a}{i\Delta\omega} + \frac{b + ci}{(1 + i\Delta\omega\tau)}\right| \tag{6.17}$$

By applying the least mean square method to the experimental result, we obtain $\tau = 35\,\mathrm{fs}$, $a = 5.21 \times 10^4\,\mathrm{m^2/V^2s}$, $b = 3.04 \times 10^{-16}\,\mathrm{m^2/V^2}$, and $c = -2.67 \times 10^{-16}\,\mathrm{m^2/V^2}$. A dephasing time of a few tens of fs agrees with the dephasing time due to electron–electron scattering as shown by Equations (6.11)–(6.16). Although we cannot determine the absolute value of the phase of $|\chi^{(3)}|$, the SHB effect has the same phase component and $\pi/2$ shifted phase component relative to the phase of the total CDP effect and CH effect. In negative detuning, the calculated line increases above 10^{13} rad/s despite a monotonous decrease in the experimental data. We cannot explain this discrepancy. An adequate explanation may require knowledge of other effects such as two-photon absorption, nonlinear effects in the clad layers, and the carrier dynamics of MQWs such as carrier capture time and carrier escape.

6.3.3 Symmetric $\chi^{(3)}$ in Quantum Dot SOAs

The asymmetry of conversion efficiency between positive and negative detuning is one of the problems in wavelength conversion based on FWM in bulk and QW (quantum well) SOAs. The use of quantum-dot (QD) as the active region in SOAs is a new technical breakthrough for

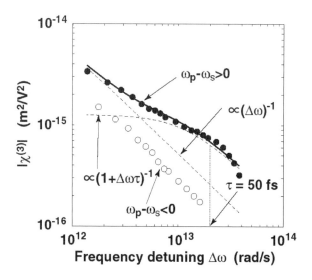

Figure 6.5 The measured $\chi^{(3)}$ by four-wave mixing in a DFB LD [38]. (Reproduced by permission of © 1997 IEEE)

symmetric wavelength conversion. In ideal QD active layers, gain peak wavelength does not change with carrier density; in other words, the change of refractive index becomes zero. The line-width enhancement factor α_c is expected to be zero as in Equation (6.3). As expected from Equation (6.2), the real number part is dominant for the CDP effect in the case of QDs. The interference between the CDP effect and the SHB effect, where the imaginary number part is dominant as in Figure 6.3, is suppressed. The conversion efficiency becomes symmetrical for detuning. Figure 6.6 shows the conversion efficiency of GaAs based QD SOAs [39]. Compared

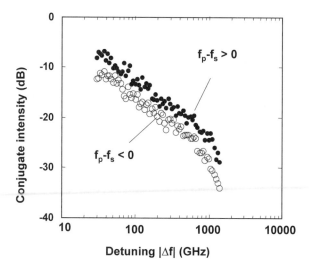

Figure 6.6 The conversion efficiency of GaAs-based QD SOAs [39]. (Reproduced by permission of © 2002 IEEE)

with the case of QWs in Figure 6.5, the difference between positive and negative detuning is small. The symmetric wavelength conversion is experimentally demonstrated.

6.4 Wavelength Conversion of Short Pulses Using FWM in Semiconductor Devices

The wavelength conversion of optical pulses with a pulse-width of less than several picoseconds has been investigated and demonstrated using FWM in SOAs [27] [40–42]. As described in Section 6.3, some third order nonlinear optical effects, especially the SHB effect, have faster response times than several hundred femtoseconds, so devices using FWM may be used as wavelength converters for ultra-fast OTDM systems of more than 1 Tbps.

There are several reports discussing the dynamics of FWM in SOAs [43–45]. In this section, the picosecond pulse response of wavelength conversion using FWM is discussed, especially the case of DFB LDs, i.e., the lasing mode of DFB LD is used for pumping. Compared with the FWM in SOAs, the FWM in a DFB LD has two advantages for short-pulse wavelength conversion. One is the effect of the stop band due to the grating structure, and the other is small gain depletion by the input pulse due to clamping of the gain by the lasing mode. We have examined these two effects using numerical simulation. The discussion can easily be applied to SOAs.

6.4.1 Model

The model of simulation of wavelength conversion in a DFB LD is shown in Figure 6.7. The electric field of the signal beam propagating to the right hand side is not affected by corrugation because its wavelength is outside the stop band [38]. Similarly to the case of SOAs, the development of the input signal is described as:

$$\frac{dE_s}{dz} = \frac{\Gamma g(\omega_s)}{2} E_s \tag{6.18}$$

where E_s is the electric field of the signal, z is the position in the propagating direction, and g is optical gain. The conjugate beam propagating to the right hand side is generated from the pump beam propagating to the right hand side, so we approximate the development of the conjugate beam as [38]:

$$\frac{dE_i}{dz} = \Gamma \left(\frac{g(\omega_i)}{2} E_i + i \frac{3\mu_0 \varepsilon_0 \omega_i^2}{8k_i} \chi^{(3)}(\omega_i; \omega_p, \omega_p, -\omega_s) R^2 E_s \right) \tag{6.19}$$

Assuming that the short optical signal pulse $E_s(0, t)e^{i\omega_s t}$ is taken from the left side, the pulse can be decomposed into frequency components by Fourier transform to give:

$$\overline{E}_s(z, \Delta\omega) = \frac{1}{\sqrt{2\pi}} \int_{-\infty}^{\infty} E_s(z, t) e^{i\Delta\omega t} dt \tag{6.20}$$

and

$$E_s(z, t) = \frac{1}{\sqrt{2\pi}} \int_{-\infty}^{\infty} \overline{E}_s(z, \Delta\omega) e^{-i\Delta\omega t} d\Delta\omega \qquad (6.21)$$

where $\Delta\omega$ is the difference from the center optical frequency of the signal. The frequency component of signal pulses propagating to the right direction given by:

$$\frac{d\overline{E}_s(z, \Delta\omega)}{dz} = -\frac{i\omega}{c_0} \eta(z) \overline{E}_s(z, \Delta\omega) - \frac{i\omega}{c_0} \frac{1}{\sqrt{2\pi}} \int_{-\infty}^{\infty} \Delta\eta(z, \Delta\omega') \overline{E}_s(z, \Delta\omega + \Delta\omega') d\Delta\omega' \qquad (6.22)$$

The frequency component of the generated conjugate pulses develops as:

$$\frac{d\overline{E}_i(z, -\Delta\omega)}{dz} = -\frac{i\omega}{c_0} \left(\eta(z) \overline{E}_i(z, -\Delta\omega) + \frac{1}{\sqrt{2\pi}} \int_{-\infty}^{\infty} \Delta\eta(z, \Delta\omega') \overline{E}_i(z, \Delta\omega + \Delta\omega') d\Delta\omega' \right)$$

$$+ i\frac{3\mu_0\varepsilon_0\omega_i^2}{8k_i} \frac{1}{2\pi} \iint \chi^{(3)} \left(\omega_i - \Delta\omega; \omega_p + \Delta\omega'', \omega_p + \Delta\omega' - \Delta\omega'', -\omega_s - \Delta\omega' - \Delta\omega \right)$$

$$\times R(\Delta\omega'') R(\Delta\omega' - \Delta\omega'') \overline{E}_s(z, \Delta\omega + \Delta\omega') d\Delta\omega'' d\Delta\omega' \qquad (6.23)$$

where $\Delta\eta$ is the change of refractive index. Input signal pulses cause a change in the carrier density. Using the rate equation, this change ΔN is described by:

$$\Delta N(z, \Delta\omega) = \frac{G(N) \cdot P_s(z, \Delta\omega)}{i\Delta\omega - \frac{dG}{dN} P_0 - \frac{1}{\tau_r}} \qquad (6.24)$$

where P_0 is the power of the pump wave, τ_r is the carrier recombination time, and P_s is frequency component of signal pulses given by:

$$P_s(z, \Delta\omega) = \frac{1}{\sqrt{2\pi}} \int_{-\infty}^{\infty} P_s(z, t) e^{i\Delta\omega t} dt \qquad (6.25)$$

The change causes the change of refractive index $\Delta\eta$:

$$\Delta\eta(z, \Delta\omega) = \frac{d\eta}{dN} \Delta N(z, \Delta\omega) \qquad (6.26)$$

The real number part of $\Delta\eta$ represents the change of refractive index, and the imaginary number part represents the change in the gain. This causes self-phase modulation (SPM) and self-gain modulation (SGM) for signal pulses, with cross-phase modulation and cross-gain modulation for the conjugate pulse.

The depletion of the carrier density gives effect not only to the signal pulses and the conjugate pulses directly, but also to the pump wave, which gives effects to the generation of

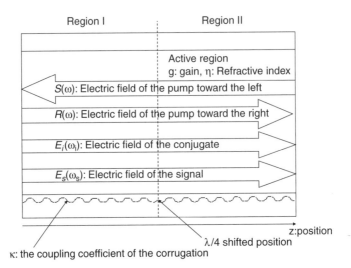

Figure 6.7 The model of four-wave mixing in a DFB LD

conjugate pulses. The pump wave profile in the $\lambda/4$ phase-shifted DFB LD at the steady state is described by:

$$-\frac{dR_I}{dz} + \left(\alpha - i\frac{\eta_R(\omega - \omega_0)}{c}\right) R_I = i\kappa S_I \tag{6.27a}$$

$$-\frac{dR_{II}}{dz} + \left(\alpha - i\frac{\eta_R(\omega - \omega_0)}{c}\right) R_{II} = i\exp(-i\phi)\kappa S_{II} \tag{6.27b}$$

$$\frac{dS_I}{dz} + \left(\alpha - i\frac{\eta_R(\omega - \omega_0)}{c}\right) S_I = i\kappa R_I \tag{6.27c}$$

$$\frac{dS_{II}}{dz} + \left(\alpha - i\frac{\eta(\omega - \omega_0)}{c}\right) S_{II} = i\exp(i\phi)\kappa R_{II} \tag{6.27d}$$

The changes of pump waves by the input pulses are given by:

$$-\frac{d\overline{R}_I(z, \Delta\omega)}{dz} + \left(\alpha - i\frac{\eta_R\Delta\omega}{c}\right) \overline{R}_I(z, \Delta\omega) + \frac{1}{\sqrt{2\pi}}\left(\Delta\alpha(z, \Delta\omega) - \frac{i\Delta\omega}{c}\Delta\eta_R(z, \Delta\omega)\right)$$

$$\times R_I(z) = i\kappa\overline{S}_I(z, \Delta\omega) \tag{6.28a}$$

$$-\frac{d\overline{R}_{II}(z, \Delta\omega)}{dz} + \left(\alpha - i\frac{\eta_R\Delta\omega}{c}\right) \overline{R}_{II}(z, \Delta\omega) + \frac{1}{\sqrt{2\pi}}\left(\Delta\alpha(z, \Delta\omega) - \frac{i\Delta\omega}{c}\Delta\eta_R(z, \Delta\omega)\right)$$

$$\times R_{II}(z) = i\exp(-i\phi)\kappa\overline{S}_{II}(z, \Delta\omega) \tag{6.28b}$$

$$\frac{d\overline{S}_I(z, \Delta\omega)}{dz} + \left(\alpha - i\frac{\eta_R\Delta\omega}{c}\right) \overline{S}_I(z, \Delta\omega) + \frac{1}{\sqrt{2\pi}}\left(\Delta\alpha(z, \Delta\omega) - \frac{i\Delta\omega}{c}\Delta\eta_R(z, \Delta\omega)\right)$$

$$\times S_I(z) = i\kappa\overline{R}_I(z, \Delta\omega) \tag{6.28c}$$

and

$$\frac{d\bar{S}_{II}(z, \Delta\omega)}{dz} + \left(\alpha - i\frac{\eta_R \Delta\omega}{c}\right)\bar{S}_{II}(z, \Delta\omega) + \frac{1}{\sqrt{2\pi}}\left(\Delta\alpha(z, \Delta\omega) - \frac{i\Delta\omega}{c}\Delta\eta_R(z, \Delta\omega)\right)$$

$$\times R_{II}(z) = i\exp(i\phi)\kappa\bar{R}_{II}(z, \Delta\omega) \qquad (6.28d)$$

The propagation of signal pulses, and conjugate pulses and spectra are evaluated using numerical simulation. Input pulses are decomposed to the frequency components using Equation (6.20). Each frequency component is calculated using Equations (6.22)–(6.28). The pulse shape of the generated conjugate and amplified signal is recomposed using Equation (6.21). The effect of detuning the spectrum of $\chi^{(3)}$ can be easily considered using this method. The filtering effect of the stop band of the corrugation in DFB can be also considered. Parameters of devices and condition are shown in Table 6.2. The spectrum of $\chi^{(3)}$ is given by Equation (6.17).

Table 6.2   Parameters for the simulation of short pulse wavelength conversion in DFB lasers

DFB laser condition		
Cavity length	1.0	mm
Pump power (Lasing power)	40.0	mW (CW)
Dispersion of refractive index	-2.1×10^{-3}	sec/m^2
$\chi^{(3)}$ in active region (parameters of equation (6.17))		
τ	35.0	fs
a	5.21×10^{-4}	m^2/V^2-s
b	3.04×10^{-16}	m^2/V^2
c	-2.67×10^{-16}	m^2/V^2
Input pulse condition		
Repetition rate	16.9	GHz (60 ps)
Pulse width	0.81	ps (Gaussian)
Peak intensity	$40 \sim 1000$	mW
Detuning frequency	500	GHz

6.4.2 The Effect of the Stop Band in DFB-LDs

At first, we discuss the effect of the stop band. In a DFB LD, the optical frequency components near the pump frequency do not propagate due to the diffraction of the grating. For example, a $\lambda/4$-shifted DFB LD with a grating coupling coefficient κ of 12.2 cm^{-1} has a stop-band width of 1.24 nm, which corresponds to 1×10^{12} rad/s. As shown in Figure 6.4, the CDP effect is dominant for detuning of less than 2×10^{12} rad/s. Slow response due to the CDP effect can be suppressed by removing the small detuning components. Figure 6.8 shows the simulated pulse shape of an input signal and its output conjugate with and without the stop-band effect. The Full Width Half Maximum (FWHM) pulse widths of the input with output conjugate with the stop-band effect, and the output conjugate without the stop-band effect are 0.85 ps,

1.2 ps, and 1.0 ps respectively. The case without the stop-band effect has a long tailing due to the CDP effect. The decreasing time of the long tailing is 10 ps, corresponding to the carrier recombination time. On the other hand, the tailing is not observed in the case with the stop band effect. The CDP effect is suppressed and fast effects, such as the SHB effect, generate the conjugate. Although the conversion efficiency from the input signal pulse to the conjugate is smaller than that without the stop-band effect and the pulse width becomes 40 % wider than that of the input, which is caused by the removal of high frequency components in the pulse, the shape of the conjugate remains close to the input. This result can be applied not only to DFB LDs but also to SOAs. In the case of SOAs, the slow component can be removed by optical filters.

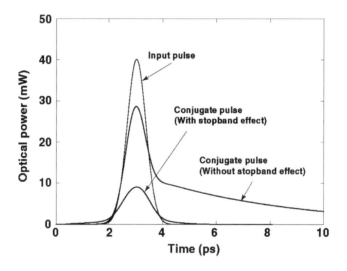

Figure 6.8 The simulated pulse shape of an input signal and its output conjugate with and without the stop band effect [46]. (Reproduced by permission of © 2000 VDE VERLAG GMBH)

6.4.3 The Effect of the Depletion of Gain

The effect of the depletion of gain due to input pulses is also examined by the simulation. As the lasing beam is generated by the gain balancing with the cavity loss, the change of gain modulates the pump beam. Figure 6.9 shows that the gain and the pump power change at the output edge of the laser. The result is shown for two peak powers of the input pulses, 40 mW and 1 W, respectively. The lasing power is 40 mW. The time of 0 ps is when an optical pulse is put at the input edge. At 10.5 ps, the amplified input pulse and the converted conjugate go out from the output edge. The changes of gain and pump are negligible when the peak input power is 40 mW. As the duty of pulse is about 5 %, the average input power is 2 mW, one order of magnitude smaller than the lasing power. This small gain change is because the gain is clamped at the threshold gain in the case of LDs. When the input power is 1 W, which is 25-times larger than the pump power, the average input power is almost same as the lasing power. In this case, depletion of gain and oscillation of pump occur. The gain at the output edge decreases rapidly by 60 % when the amplified pulse goes out from the output edge. The recovery time of the gain is 10 ps, which corresponds to the carrier recombination time. On the other hand, the change of the pump is only 15 %. This stability of the pump is due to the cavity of the laser, which functions as a reservoir of the pump.

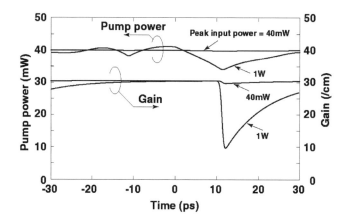

Figure 6.9   The gain changes and pump changes at the output edge of the laser [46]. (Reproduced by permission of © 2000 VDE VERLAG GMBH)

6.4.4 The Pulse Width Broadening in FWM Wavelength Conversion

Figure 6.10 shows the input signal pulses and the conjugate when the input pulse is 1 W. For comparison, the case without the saturation effect of the gain and pump in Figure 6.9 is also

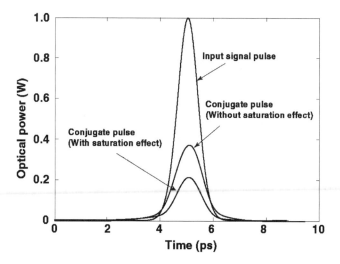

Figure 6.10   The results of simulation of input signal pulses and the conjugate [46]. (Reproduced by permission of © 2000 VDE VERLAG GMBH)

shown. The conjugate decreases due to the decrease of the pump, compared with the case without the depletion of gain and pump. The decrease of pulse intensity is 45 %, 30 % being due to the depletion of the pump, because the intensity of the conjugate is proportional to the square of the pump, and the remaining 15 % is due to the depletion of gain. The FWHM pulse width of the input, the conjugate without the saturation effect, and the conjugates with saturation effect are 0.85 ps, 1.2 ps, and 1.1 ps respectively. The reason for the broadening of the pulse width is the stop-band effect as discussed before. The pulse width with the saturation effect is narrower than that without the saturation effect. This is because the decrease of the carrier density causing the saturation of the gain causes a decrease of refractive index. The rear part of the pulse propagates faster than the front part, due to the decrease of the refractive index.

The simulations show that the wavelength conversion of picosecond pulses is possible by FWM in DFB-LDs. The slow component of third-order nonlinear effects such as the CDP effect, is suppressed by wavelength filters such as the stop band of corrugation of DFB-LDs.

6.5 Experimental Results of Wavelength Conversion Using FWM in SOAs or LDs

In this section, several experimental results for wavelength conversion using FWM in semiconductor optical devices are presented. Special characteristics of FWM, such as short pulse wavelength conversion, OTDM signal conversion, format-free conversion, and the generation of conjugate signals, are demonstrated.

6.5.1 Wavelength Conversion of Short Pulses Using a DFB-LD

The first example is the wavelength conversion of short-pulses. The wavelength conversion of subpicosecond short pulses is required in ultrafast OTDM systems of more than 1 Tbps. As discussed in Section 6.4, the wavelength conversion of such short-pulses is possible by FWM in DFB-LDs.

The arrangement of experiments using DFB-LDs [46] is shown in Figure 6.11. The signal pulses were generated by an OPO (optical parametric oscillator). Their pulse width and repetition rate were 200 fs and 100 MHz respectively. Their selected wavelength was 1558.0 nm. We used two optical BPFs (band path filters) with 2.64 nm and 2.61 nm bandwidths for pulse shaping. The peak power of the pulse was estimated to be 0.7 W. The DFB LD for wavelength conversion was a $\lambda/4$ shifted DFB LD with length of 900 μm. The lasing wavelength was 1553 nm and the lasing power was 17 mW. The detuning between the input signal and the pump was 5.0 nm. The output conjugate was selected by a BPF with a bandwidth of 2.66 nm, and amplified by two Er-doped fiber amplifiers (EDFAs). The pulse shape of the conjugate was measured by an autocorrelator.

The autocorrelation traces of the input signal and the converted conjugate are shown in the inset. When a hyperbolic cosine pulse shape is assumed, the pulse widths were estimated to be 0.61 ps and 0.98 ps respectively. The reason for the broadening of 60 % of the conjugate pulse is not clear. The broadening of 40 % can be explained by the stop-band effect as discussed in Section 6.4. The narrowing of over 10 % due to the saturation effect occurs because the peak power is 40-times larger than the lasing power. The broadening of 30 % may be due to the

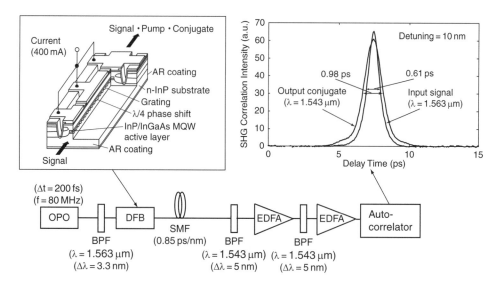

Figure 6.11 The arrangement of the experiment on the wavelength conversion of short-pulses using DFB-LDs [46]. The right inset shows the auto-correlation traces of the input signal, the output signal and the conjugate. (Reproduced by permission of © 2000 VDE VERLAG GMBH)

difference in the dispersion in the measurement systems between signal and conjugate, which does not appear in the simulation. The experiment shows that the wavelength conversion of a pulse of a few picoseconds is possible by using FWM in DFB LDs.

6.5.2 *Wavelength Conversion of 160-Gb/s OTDM Signal Using a Quantum Dot SOAs*

The other example of wavelength conversion is that of a 160-Gb/s OTDM signal using QD SOAs [47]. An InP-based QD-SOA with five InAs quantum-dot layers, which has an optical gain in the 1.5-μm wavelength range, was used. A 160-Gb/s RZ-PRBS signal is generated by optical multiplexing of a short pulse from the fiber ring laser with a center wavelength of 1550 nm and a repetition rate of 10 GHz. The power of the signal light is −1.6 dBm. The pumping wavelength and power are 1545 nm and 14 dBm, respectively. The wavelength-converted signal travels through a band-pass filter and is demultiplexed down to 10 Gb/s. Figure 6.12 shows the experimental eye-patterns of the signals before and after wavelength conversion by the QD-SOA. The spectrum before eliminating the signal and pumping light is also shown. Although the optical signal-to-noise ratio of the wavelength-converted signal is about 20 dB, and the measured eye-pattern after wavelength conversion is still noisy, we succeed in converting the wavelength of a 160-Gb/s signal using QD-SOA.

Compared with GaAs-based QD-SOAs for the 1.3-μm wavelength range, the wavelength conversion characteristics of InP-based QD-SOAs is as yet unsatisfactory. The size of InP-based QDs is larger than that of GaAs-based QDs. Energy level intervals in InP-based

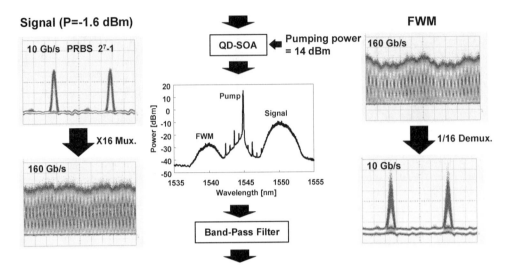

Figure 6.12   The experiment on wavelength conversion of 160-Gbps signals by an InP-based 1.5 μm QD-SOA

QDs are small. As a result, the symmetric conversion efficiency for the detuning observed in GaAs-based QD-SOAs as described in Section 6.3 has not been confirmed for the case of InP-based SOAs. For realistic application in photonic network systems, the wavelength conversion in the 1.55 μm wavelength range is important. Improvement of InP-based QD structure is necessary.

6.5.3 Format-free Wavelength Conversion

The third example is format-free wavelength conversion. Format-free wavelength conversion is one of the outstanding merits of FWM, because not only the amplitude information but also the phase information of the input signals can be converted using this scheme. However, the sign of the optical phase is inverted when the signals are converted to the conjugate. Complete format-free wavelength conversion cannot be realized by wavelength conversion to the conjugate. On the other hand, the wavelength conversion to the replica, where the sign of the optical phase is same before and after conversion, can achieve complete format-free wavelength conversion.

The wavelength conversion of OOK and DPSK signals to replicas was demonstrated using two nondegenerate pumps. An MZI-SOA was used for this experiment [48]. Figure 6.13 shows the experimental setup. The 4-mm long MZI-SOA consists of two 1.5-mm long SOAs with a strained bulk active layer and 1:1 MMI (multi mode interference) couplers. The injected current to two SOAs is 400 mA each and the peak wavelength of amplified spontaneous emission (ASE) spectrum is 1510 nm. The unsaturated gain and saturation power under a driving current of 400 mA are 24 dB and 12 dBm, respectively. Pumping lights (P1 and P2) are guided into Port A and a signal (S) is injected to Port B. The input signal is modulated at 10 Gb/s using a LiNbO$_3$ (LN) modulator with a $2^7 - 1$ pseudorandom binary sequence. The NRZ-OOK signal is generated when the LN modulator is driven with V_π and the NRZ-DPSK signal is generated when driven with $2V_\pi$. The converted OOK signal is received directly. The converted DPSK signal is decoded as the intensity-modulated electric signal with delayed interferometer and balanced detector. The phase-conjugated lights and an unconjugated replica of the signal come out from Port C with the pumping lights suppressed. The replica is filtered out using band-pass filters and a 25–50 GHz interleaver, with its passband aligned to the ITU grid of the signal.

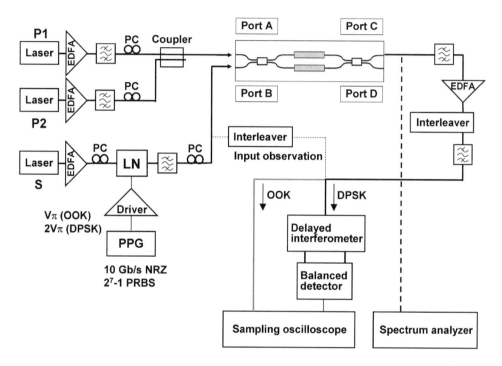

Figure 6.13   Experimental setup of format free wavelength conversion to replicas of OOK and DPSK signals [48]. (Reproduced by permission of © 2006 VDE VERLAG GMBH)

Figure 6.14 shows the spectrum of direct output from Port C and eye patterns of the converted replica for NRZ-OOK and NRZ-DPSK modulation. It is fair to compare the eye patterns with those of input signals through the interleaver, because the converted outputs are also extracted by it. In both modulation formats, the conditions are exactly the same, and are as follows. The input peak power of signal (S), pump 1 (P1), and pump 2 (P2) are −6.4, 3.8, and −6.4 dBm, respectively. The wavelength of the input is 1546.12 nm and that of the converted replica is 1562.23 nm. The wavelength of P1 is 1546.24 nm and the S-P1 detuning is set at −15 GHz (−0.12 nm). The conversion efficiency is −6.0 dB. Comparing intensities of the output pumping lights from Ports C and D, the suppression ratio of P2 is 28 dB. By using MZI configuration, the pump is well suppressed. Clear eye openings are observed in the converted replica not only in the NRZ-OOK format but also in the DPSK. The method described here can convert the wavelength of the signals to their replicas and is applicable to different modulation formats using the same operation conditions. This indicates that this method should achieve format-free and completely transparent wavelength conversion.

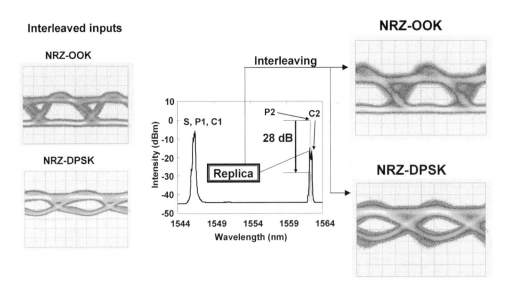

Figure 6.14   The spectrum of direct output from port C and eye patterns of the converted replica for NRZ-OOK and NRZ-DPSK modulation [48]. (Reproduced by permission of © 2006 VDE VERLAG GMBH)

6.5.4 *Chromatic Dispersion Compensation of Optical Fibers Using FWM in DFB-LDs*

The final example is the application of FWM for chromatic dispersion compensation of optical fibers [49, 50]. As the sign of the optical phase is inverted when signals are converted to the optical conjugate, the change of the optical phase due to chromatic dispersion in optical fiber is opposite to the change of the conjugate with the same dispersion. Waveform deteriorated by chromatic dispersion can be restored by converting signals to optical conjugates and transmitting the optical conjugate in an optical fiber with the same dispersion. Not only the first order of dispersion but also higher odd orders of dispersion can be compensated for by using optical phase conjugate. In addition, the nonlinear effect of optical fibers such as SPM, and XPM are also compensated.

A polarization-insensitive phase conjugate wave generator using bidirectional FWM in DFB-LD has been proposed and demonstrated [49]. Figure 6.15 shows an example of the experiment for the dispersion compensation of optical fibers for 40 Gbps signals using the phase conjugate wave generator. A signal of 40 Gbps deteriorates after transmission in a 50-km optical fiber with a dispersion of +18.2 ps/nm/km. The deteriorated signal is input to an optical circulator (OC) followed by a polarization beam splitter (PBS) through which the signal is divided into two orthogonally polarized components, a transmitted x-component and a reflected y-component. The x-component (counter-clockwise) with the same polarization as that of the DFB-LD, is incident on the first facet of the DFB-LD and converted to a conjugate. Conversely, the polarization state of the y-component (clockwise) is converted to the orthogonal, which is the same as that of the x-component, before being incident on the second facet of the DFB-LD. The wave is then converted to a conjugate. The two converted waves are coupled through a PBS and output from the third port of the OC. When the injection

currents in three electrodes are controlled to give the same conversion efficiencies for the
x-component and the y-component, polarization-insensitive conversion can be realized using
the above configuration. The phase-conjugate wave was then transmitted through the second
fiber (50 km) with a dispersion of +17.8 p/nm/km. The waveform distortion was compensated,
completely.

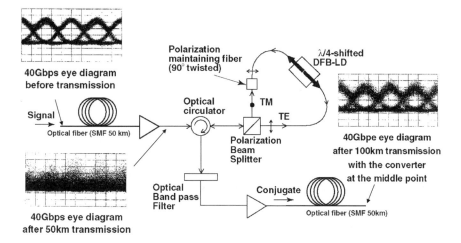

Figure 6.15 The dispersion compensation of optical fibers by the polarization-insensitive phase
conjugate wave generator using FWM in DFB-laser

6.6 The Future View of Wavelength Conversion Using FWM

In future photonic network systems, wavelength conversion devices will play important roles.
In particular, they are indispensable in reconfigurable optical add-drop multiplexing (ROADM)
systems or optical cross connect (OXC) systems. In future network systems, various format signals will be used. Because of the possibility of transparent wavelength conversion for various
formats, coherent type wavelength converters have huge potential. Among them, wavelength
converters using FWM in SOAs or LDs have largest potential because of their high conversion efficiency and their small size. For practical uses, several points of improvement are
required.

One of these is an improvement in the optical signal-to-noise ratio (OSNR) or the noise
figure (NF). Although high conversion efficiencies over unity are possible, due to optical gain
for input signals and converted signals, the degradation of OSNR or NF is caused by amplified
spontaneous emission (ASE). The improvement of intrinsic conversion efficiency with low
optical gain is required. For this purpose, the design of the device structure to increase the
optical confinement and the optimization of the interaction length should be considered. The
exploitation of materials for the active region to increase $\chi^{(3)}$ is should also be considered.

Realization of a higher efficiency conversion over the whole C-band is also required. The
use of QD structure in the active region is effective in suppressing the asymmetry of conversion
for negative and positive detuning. The wavelength conversion characteristics of InP-based

QD-SOAs have not reached the theoretically expected ones. The improvement of InP-based QDs covering the 1.5-μm wavelength range is required. Detuning dependence, however, exists even in the case of the QDs. The conversion efficiency becomes low when a signal of a wavelength at the end of the C-band is converted to a signal at the wavelength of the other end. The use of two degenerate pumps is a method for realizing the flat conversion efficiency over C-band [51].

To decrease the size of wavelength converters, integration with a pump light source is promising. Integration with a wavelength-tunable light source and array integration of wavelength converters for wavelength cross-connection of WDM signals will be necessary. Similarly, integration with tunable optical filters and other optical components is also desired for decreasing the size.

The power consumption is the other important issue. The increase of the power consumption of routers will be a social problem in the future. For the wavelength converters in future photonic network systems, the reduction of the power consumption will be also indispensable. The increase of the conversion efficiency by a change in the device structure for an increase in the optical confinement and interaction length is effective for the reduction of the power. In addition, the design of optical circuit reusing pumps may be considered.

The polarization dependence of conversion efficiency is an inherent problem in coherent type wavelength converters such as FWM wavelength converters. Polarization diversity systems as shown in Section 6.5 may be a method of solving the problem. The other solution is the use of two pumps with orthogonal polarizations.

When the problems mentioned above can be solved, the performances of wavelength converters using FWM in semiconductor devices could reach a practicable level. Then, the wavelength converters using FWM will have a great impact on photonic network systems.

6.7 Summary

Wavelength conversion schemes were reviewed. As one of the promising candidates, wavelength conversions using four-wave mixing in SOAs or LDs were presented. The principle of FWM in LDs or SOAs was also discussed. The physics of wavelength conversion of subpicosecond pulses was also described. The wavelength conversion of short pulses or 160-Gb/s OTDM signals, format-free wavelength conversion, and application of the conjugate wave were demonstrated experimentally. The future view of wavelength conversion using FWM was also discussed. Wavelength conversion is an indispensable function in future photonic network systems. Especially, wavelength conversion using FWM should play an important rule in such systems.

References

[1] S. J. B. Yoo, 'Wavelength conversion technologies for WDM network applications,' *IEEE J. Lightwave Technol.*, **14**, 955–966 ((1996)).

[2] H. Kuwatsuka, 'Four-wave mixing in semiconductor medium for optical network,' *Technical Digest of the 4th Pacific Rim Conference on Lasers and Electro-Optics*(CLEO Pacific Rim '01), Volume1, pp.I-442–I-443, Chiba (Japan) (2001).

[3] M. Nakazawa, 'Tb/s OTDM Technology,' *Proceedings 27th European Conference on Optical Communication*, Amsterdam, Tu.L.2.3 (2001).

[4] S. Ono, R. Okabe, F. Futami, and S. Watanabe, 'Novel demultiplexer for ultra high speed pulses using a perfect phase-matched parametric amplifier,' *Proceedings Optical Fiber Communication Conference (2006) and the (2006) National Fiber Optic Engineers Conference*, OWT2 (2006).

[5] T. Sakamoto, F. Futami, K. Kikuchi, S. Takeda, Y. Sugaya, and S. Watanabe, 'All-optical wavelength conversion of 500-fs pulse trains by using a nonlinear-optical loop mirror composed of a highly nonlinear DSF,' *IEEE Photon. Technol. Lett.*, **13**, 502–504 (2001)

[6] S. Watanabe and F. Futami, 'Optical Signal Processing Using Nonlinear Fibers,' *Proceedings 28th European Conference on Optical Communication* (ECOC (2002).), Volume 3, pp.1–2 (2002).

[7] M. L. Nielsen, M. Nord, M. M.N. Petersen, B. Dagens, A. Labrousse, A. R. Brenot, B. Martin, S. Squedin, and M. Renaud, '40 Gbit/s standard-mode wavelength conversion in all-active MZI with very fast response,' *Electron. Lett.* **39**, 385–386 (2003)

[8] M. L. Nielsen and J. Mork, 'Bandwidth enhancement of SOA-based switches using optical filtering: theory and experiment,' *ECOC (2005) Proceedings*, Volume 2, Tu3.5.7, pp.237–238 (2005).

[9] H. Ishikawa, 'Ultrafast all-optical switches for signal processing beyond 100Gb/s', *Proceedings Pacific Rim Conference on Lasers and Electro-Optics*, pp.1275–1276 (2005).

[10] S. Nakamura, T. Tamanuki, Y. Ueno, K. Tajima, 'Ultrafast optical demultiplexing, regeneration, and wavelength-conversion with symmetric-Mach–Zehnder all-optical switches', *Proceedings Optical Fiber Communications Conference*, Volume 1, pp. 346–347 (2003).

[11] Hodeok Jang, Sub Hur, Yonghoon Kim, and Jichai Jeong, 'Theoretical investigation of optical wavelength conversion techniques for DPSK modulation formats using FWM in SOAs and frequency comb in 10 Gb/s transmission systems,' *J. Lightwave Technol.*, **23**, 2638–2646 (2005).

[12] Yong Wang, and Chang-Qing Xu, 'Analysis of ultrafast all-optical OTDM demultiplexing based on cascaded wavelength conversion in PPLN waveguides,' *IEEE Photon. Technol. Lett.*, **19**, 495–497 (2007).

[13] R. Ludwig and G. Raybon, 'BER measurements of frequency converted signals using four-wave mixing in a semiconductor laser amplifier at 1, 2.5, 5 and 10 Gbit/s,' *Electron. Lett.*, **30**, 338–339 (1994).

[14] J. Zhou, N. Park, K. J. Vahala, M. A. Newkirk, and B. I. Miller, 'Four-wave mixing wavelength conversion efficiency in semiconductor traveling-wave amplifiers measured to 65 nm of wavelength shift,' *IEEE Photon. Technol. Lett.*, **6**, 984–987 (1994).

[15] A. D'Ottavi, E. Iannone, A. Mecozzi, S. Scotti, and P. Spano, '4.3 terahertz four-wave mixing spectroscopy of InGaAsP semiconductor amplifiers,' *Appl. Phys. Lett.*, **65**, 2633–2635 (1994).

[16] A. Uskov, J. Mork, J. Mark, M. C. Tatham and G. Sherlock, 'THz four-wave mixing in semiconductor optical amplifiers,' *Appl. Phys. Lett.*, **65**, 944–946 (1994).

[17] A. D'Ottavi, E. Iannone, A. Mecozzi, S. Scotti, and P. Spano, 'Investigation of carrier heating and spectral hole burning in semiconductor amplifiers by highly nondegenerate four-wave mixing,' *Appl. Phys. Lett.*, **64**, 2492–2494 (1994).

[18] M. A. Newkirk and B. I. Miller, 'Terahertz four-wave mixing spectroscopy for study of ultrafast dynamics in a semiconductor optical amplifier,' *Appl. Phys. Lett.*, **63**, 1179–1181 (1993).

[19] J. Zhou, N. Park, J. W. Dawson, Kerry J. Vahala, M. A. Newkirk and B. I. Miller, 'Efficiency of broadband four-wave mixing wavelength conversion using semiconductor traveling-wave amplifiers,' *IEEE Photon. Technol. Lett.*, **6**, 50–52 (1994).

[20] R. Schnabel, U. Hilbk, Th. Hermes, P. Meissner, C. Helmolt, K. Magari, F. Raub, W. Pieper, F. J. Westphal, R. Ludwig, L. Kuller, and H. G. Weber, 'Polarization insensitive frequency conversion of a 10-channel OFDM signal using four-wave-mixing in a semiconductor laser amplifier,' *IEEE Photon. Technol. Lett.*, **6**, 56–58 (1994).

[21] M. C. Tatham, G. Sherlock, and L. D. Westrook, '20-nm optical wavelength conversion using nondegenerate four-wave mixing,' *IEEE Photon. Technol. Lett.*, **5**, 1303–1306 (1993).

[22] A. D'Ottavi, F. Martelli, P. Spano, A. Mecozzi, S. Scotti, R. Dall'Ara, J. Eckner, and G. Guekos, 'Very high efficiency four-wave mixing in a single semiconductor traveling-wave amplifier,' *Appl. Phys. Lett.*, **68**, 2186–2188 (1996).

[23] F.Girardin, J.Eckner, G. Guekos, R. Dall'Ara, A. Mecozzi, A. D'Ottavi, F. Martelli, S. Scotti and P. Spano, 'Low-noise and very high-efficiency four-wave mixing in 1.5-mm-long semiconductor optical amplifiers,' *IEEE Photon. Technol. Lett.*, **9**, 746–748 (1997).

[24] A. Martelli, A. D'Ottavi, L. Graziani, A. Mecozzi, P. Spano, G. Guekos, J. Eckner, and R. Dall'Ara, 'Pump-wavelength dependence of FWM performance in semiconductor optical amplifiers,' *IEEE Photon. Technol. Lett.*, **9**, 743–7445 (1997).

[25] A. D'Ottavi, P. Spano, G. Hunziker, R. Paiella, R. Dall'Ara, G. Guekos, and K. J. Vahara, 'Wavelength conversion at 10 Gb/s by four-wave mixing over a 30-nm interval,' *IEEE Photon. Technol. Lett.*, **10**, 952–954 (1997).

[26] L. Y. Lin, J. M. Wiesenfeld, J. S. Perino, and A. H. Gnauck, 'Polarization-insensitive wavelength conversion up to 10 Gb/s based on four-wave mixing in a semiconductor optical amplifier,' *IEEE Photon. Technol. Lett.*, **10**, 955–957 (1998).

[27] J. Inoue and H. Kawaguchi, 'Time-delay characteristics of four-wave mixing among subpicosecond optical pulses in a semiconductor optical amplifier,' *IEEE Photon. Technol. Lett.*, **10**, 1566–1568 (1998).

[28] P. P. Iannone, P. R. Prucnal, G. Raybon, U. Koren, and C. A. Burrus, 'Nanometer wavelength conversion of picosecond optical pulses using cavity-enhanced highly nondegenerate four-wave mixing in semiconductor Lasers,' *IEEE J. Quantum Electron.*, **31**, 1285–1291 (1995).

[29] E. Cerboneschi, D. Hennequin, and E. Arimondo, 'Frequency conversion in external cavity semiconductor lasers exposed to optical injection,' *IEEE J. Quantum. Electron.*, **22**, 192–200 (1996).

[30] A. Mecozzi, A. D'Otavi and R. Hui, 'Nearly degenerate four-wave mixing in distributed feedback semiconductor lasers operating above threshold,' *IEEE J. Quantum. Electron.*, **29**, 1477–1487 (1993).

[31] R. Hui, S, Benedetto, and I. Montrosset, 'Optical frequency conversion using nearly degenerate four-wave mixing in a distributed-feedback semiconductor laser: theory and experiment,' *IEEE J. Lightwave Technol.*, **11**, 2026–2031 (1993).

[32] S. Murata, A. Tomita, J. Shimizu, and A. Suzuki, 'THz optical-frequency conversion of 1 Gb/s-signals using highly nondegenerate four-wave nixing in an InGaAsP semiconductor Laser,' *IEEE Photon. Technol. Lett.*, **3**, 1021–1023 (1991).

[33] S. Murata, A. Tomita, J. Shimizu, M. Kitamura, and A. Suzuki, 'Observation of highly nondegenerate four-wave mixing in an InGaAsP multi quantum well Laser,' *Appl. Phys. Lett.*, **58**, 1458–1460 (1991).

[34] H. Kuwatsuka, H. Shoji, M. Matsuda, and H. Ishikawa, 'THz frequency conversion using nondegenerate four-wave mixing process in a lasing long-cavity λ/4-shifted DFB laser,' *Electron. Lett.*, **31**, 2108–2110 (1995).

[35] S. Yamashita, S. Y. Set, and R. I. Laming, 'Polarization independent, all-fiber phase conjugation incorporation inline fiber DFB lasers,' *IEEE Photon. Technol. Lett.*, **10**, 1407–1409 (1998).

[36] H. Kuwatsuka, T. Simoyama, and H. Ishikawa, 'Enhancement of third-order nonlinear optical susceptibilities in compressively strained quantum wells under the population inversion condition', *IEEE J. Quantum. Electron.*, **35**, 1817–1825 (1999).

[37] S. Seki, and K. Yokoyama, 'Intrasubband scattering in highly excited semiconductor quantum wells with biaxial strain,' *Phys. Rev. B*, **50**, 1663–1670 (1994).

[38] H. Kuwatsuka, H. Shoji, M. Matsuda, and H. Ishikawa, 'Nondegenerate four-wave mixing in a long-cavity λ/4-shifed DFB laser using its lasing beam as pump beams,' *IEEE J. Quantum. Electron.*, **33**, 2002–2010 (1997).

[39] T. Akiyama, H. Kuwatsuka, N. Hatori, Y. Nakata, H. Ebe, and M. Sugawara, 'Symmetric highly efficient wavelength conversion based on four-wave mixing in quantum dot optical amplifiers,' *IEEE Photon. Technol. Lett.*, **14**, 1139–1141 (2002).

[40] S. Diez, C. Schmidt, R. Ludwig, H. G. Weber, K. Obermann, S. Kindt, I. Koltchanov, and K. Petermann, 'Four-wave mixing in semiconductor optical amplifiers for frequency conversion and fast optical switching,' *IEEE of J. Selected Topics Quantum Electron.*, **3**, 1131–1145 (1997).

[41] J. Inoue and H. Kawaguchi, 'Highly nondegenerate four-wave mixing among subpicosecond optical pulses in a semiconductor optical amplifier,' *IEEE Photon. Technol. Lett.*, **10**, 349–351 (1998).

[42] G. Berrettini, A. Simi, A. Malacarne, A. Bogoni, and L. Potí, 'Ultrafast integrable and reconfigurable XNOR, AND, NOR, and NOT photonic logic gate,' *IEEE Photon. Technol. Lett.*, **18**, 917–919 (2006).

[43] Christina (Tanya) Politi, D. Klonidis, and M. J. O'Mahony, 'Dynamic behavior of wavelength converters based on FWM in SOAs,' *IEEE J. Quantum Electron.*, **42**, 108–125 (2006).

[44] N. Kumar Das, Y. Yamayoshi, and H. Kawaguchi, 'Analysis of basic four-wave mixing characteristics in a semiconductor optical amplifier by the finite-difference beam propagation method,' *IEEE J. Quantum Electron.*, **36**, 1184–1192 (2000).

[45] N. Kumar Das, Y. Yamayoshi, T. Kawazoe, and H. Kawaguchi, 'Analysis of optical DEMUX characteristics based on four-wave mixing in semiconductor optical amplifiers,' *J. Lightwave Technol.*, **19**, 237–246 (2001).

[46] H. Kuwatsuka, T. Akiyama, B. E. Little, T. Simoyama, and H. Ishikawa, 'Wavelength conversion of picosecond optical pulses using four-wave mixing in a DFB Laser,' *Technical Digest of 26th European Conference on Optical Communication* (ECOC (2000)), Volume 3, pp.65–66 (2000).

[47] K. Otsubo, T. Akiyama, H. Kuwatsuka, N. Hatori, H. Ebe, and M. Sugawara, 'Automatically controlled C-band wavelength conversion with constant output power based on four-wave mixing in SOA's,' *IEICE Trans. Electron.*, **E88-C**, 2358–2365 (2005).

[48] K. Otsubo, S. Tanaka, S. Tomabechi, K. Morito, H. Kuwatsuka, 'Wavelength conversion to replicas of OOK and DPSK signals with one configuration based on dual pump nearly degenerated four-wave mixing in a Mach–Zehnder interferometer SOA,' *Proceedings of 32nd European Conference on Optical Communication* (ECOC '06), Volume 4, pp.173–174 (2006).

[49] S. Watanabe, H. Kuwatsuka, S. Takeda and H. Ishikawa, 'Polarization-insensitive wavelength conversion and phase conjugation using bi-directional forward four-wave mixing in a lasing DFB-LD,' *Electron. Lett.*, **33**, 316–317 (1997).

[50] A. Mecozzi, G. Contestabile, F. Martelli, L. Graziani, A. D'Ottavi, P. Spano, R. Dall'Ara, J. Eckner, F. Girardin, and G. Guekos, 'Optical spectral inversion without frequency shift by four-wave mixing using two pumps with orthogonal polarization,' *IEEE Photon. Technol. Lett.*, **10**, 355–357 (1998).

[51] Dar-Zu Hsu, San-Liang Lee, Pei-Miin Gong, Yu-Min Lin, Steven S. W. Lee, and Maria C. Yuang, 'High-efficiency wide-band SOA-based wavelength converters by using dual-pumped four-wave mixing and an assist beam,' *IEEE Photon. Technol. Lett.*, **16**, 1903–1905 (2004).

7

Summary and Future Prospects

Hiroshi Ishikawa

7.1 Introduction

In the previous chapters we have described various ultrafast signal processing devices. In this chapter we at first review ultrafast transmission experiments and then discuss the requirements and issues further to be overcome for ultrafast signal processing devices.

There are two groups of transmission experiment using devices described in this book. One comprises the experiments done at the Femtosecond Technology Research Association (FESTA, Laboratory for Femtosecond Technology Project) using mode-locked lasers (Chapter 2) and SMZ gate switches (Chapter 3). The other is a field experiment using a test bed called Japan Gigabit Network II (JGN II) using EAM based light sources (Chapter 2) in the framework of the project 'Research and Development on Ultrahigh-speed Backbone Photonic Network Technologies'. Firstly we review these experiments. Secondly we review recent transmission experiments taking examples form the ECOC (European Conference on Optical Communication) papers from 2005 to 2007.

These reviews will illustrate how devices are used in the ultrafast experimental systems and also show the technological trend for ultrafast transmission systems. Base on these reviews, we look at devices described in this book and discuss the issues further to be overcome and their prospects. We also discuss the necessity for some new devices and new technologies.

7.2 Transmission Experiments

7.2.1 FESTA Experiments

Suzuki *et al.* reported 160-Gb/s eight wavelength transmissions over 140 km using the configuration shown in Figure 7.1 at the European Conference on Optical Communication (ECOC) 2003 [1]. The total capacity is 1.28 Tb/s.

A monolithically integrated DBR mode-locked laser diode (ML-LD), described in Chapter 2, was used as the light source. The DBR ML-LD was operated at a repetition rate of 40 GHz

---

Ultrafast All-Optical Signal Processing Devices   Edited by Hiroshi Ishikawa
© 2008 John Wiley & Sons, Ltd

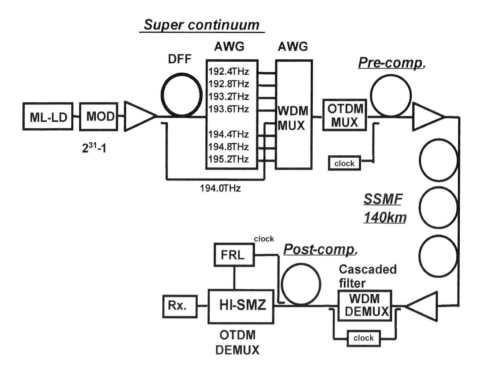

Figure 7.1 Configuration of a transmission experiment for 160-Gb/s, 8 WDM by Suzuki *et al.* [1]. (Reproduced by permission of 2003 © AEIT)

and at a wavelength of 1545.3 nm. The repetition rate was fine tuncd to 39.81312 GHz using a hybrid mode-locking operation as illustrated in Chapter 2. With this operating mode the timing jitter was suppressed to be below 200 fs. The pulse width was 2.3 ps and the spectral width was 1.6 nm. The pulses from the DBR ML-LD were modulated using a LiNbO$_3$ modulator at 40 Gb/s with PBRS (pseudo random binary sequence) of $2^{31} - 1$. After amplification by EDFA, the super continuum light was generated using a self-phase modulation effect in dispersion flat fiber (DFF). The ITU grid wavelengths of seven channels with 400-GHz spacing were sliced from the super continuum. The original modulated signal after the EDFA was also used as one of the channels to form eight WDM channels. One more AWG having a flat-top band-pass with a 3dB-width of 2.8 nm was used for wavelength multiplexing. With this AWG, coherent beat noise between the adjacent channels was suppressed. Then OTDM multiplication was carried out. The transmission line was a single span of 140-km SSMF (standard single mode fiber). Pre-compensation of the dispersion was done using dispersion compensation fiber (DCF) for the SSML span of 100 km. Post compensation was done using positive and negative dispersion slope DCF, which finely compensated the total dispersion and dispersion slope. The total dispersion and dispersion slope of the transmission line after compensation were 0.1 ps/nm and 0.1 ps/nm/nm, respectively. On the receiver side, a four-stage cascaded tunable optical filter having a bandwidth of 2.8 nm was used as a channel selector. The DEMUX to 10 Gb/s was done using an SMZ gate switch, described in Chapter 3. The control pulse for the SMZ gate switch was generated by a tunable mode-locked fiber laser. The clock to synchronize the

control pulse was transmitted separately through the transmission line. Figure 7.2(a) shows
the back-to-back bit-error rate for all eight channels, and Figure 7.2(b) shows the bit-error
rate after transmission. Although there were 2–3 dB penalties for the transmission, no floor
was observed in the error-rate curve and an error rate of 10^{-9} was achieved for all eight
channels.

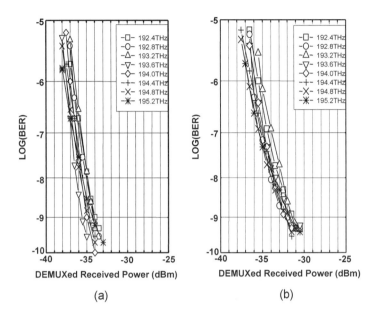

Figure 7.2   Bit-error rates for eight tributary channels. (a) Error rate for back to back. (b) Error rate
after 140 km transmission [1]. (Reproduced by permission of 2003 © AEIT)

Suzuki *et al.* also performed 320-Gb/s ten-wavelength transmission experiments over 40 km of
SSMF and reported them in ECOC2004 [2]. The total capacity was 3.2 Tb/s. The experimental
set up is shown in Figure 7.3. Again the DBR mode-locked laser was used as a 40-Gb/s
pulse source with a pulse width of 2.3 ps. To give a 320-Gb/s transmission, a dispersion
decreasing fiber (DDF) was used to compress the pulse. As there was large fluctuation in the
spectrum around the pump wavelength after DDF, additional AWG was used to reshape the
spectrum. Then a super continuum was generated using a dispersion flat fiber (DFF) with a
normal dispersion of the length of 2 km. Using AWG, ITU grid wavelengths were sliced. Pulse
width as estimated from the bandwidth of the AWG was 1.0 ps. Channels were wavelength
multiplexing for odd channels and even channels separately using second AWG, and then
modulated by EAM. The OTDM multiplexing was done using a polarization beam splitter
(PBS) to generate a polarization interleaved OTDM/WDM signal. Dispersion compensation
was achieved by combining negative slope fibers and positive slope fibers. The total dispersion
and dispersion slope of the transmission line were less than 0.05 ps/nm and 0.05 ps/nm/nm,
respectively. The DEMUX at the receiver side was again carried out using an SMZ gate switch
in a similar manner to the first experiment. The back-to-back bit-error rate curves and those after
40 km transmission are shown in Figure 7.4(a) and (b), respectively. No floor was observed up
to an error rate of 10^{-9} for all tributary channels. Through these experiments, the usefulness
of the DBR mode-locked LD and the SMZ gate switch was verified.

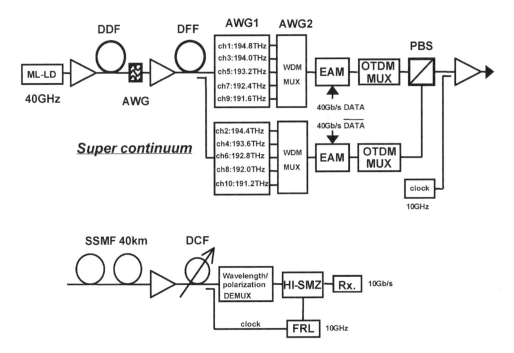

Figure 7.3 Configuration of transmission experiment for 320-Gb/s, 10 WDM by Suzuki *et al.* [2]. (Reproduced by permission of © 2004 Royal Institute of Technology (KTH))

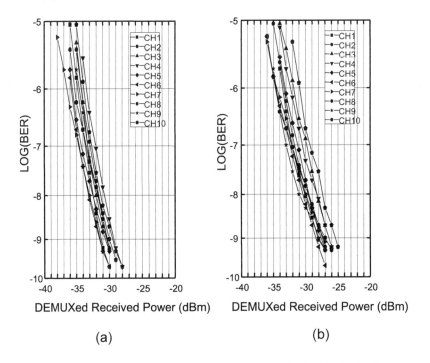

Figure 7.4 Bit-error rates for 10 tributary channels. (a) Error rate for back to back. (b) Error rate after 40 km transmission [2]. (Reproduced by permission of © 2004 Royal Institute of Technology (KTH))

7.2.2 Test Bed Field Experiment

In the project 'Research and Development on Ultrahigh-speed Backbone Photonics Network Technologies', Murai *et al.* performed a field transmission at 160 Gb/s using JGNII test bed [3]. In this experiment, EAM based light source described in Chapter 2 was used.

Figure 7.5 shows the experimental set up. The map of the location of the test bed is shown Figure 7.6. The EAM based light source described in Chapter 2 was used to generate a 160-Gb/s CS-RZ OTDM signal. The 160-Gb/s CS-RZ signal was transmitted through 63.5 km SSMF between Keihanna (NICT Laboratory) and Dojima via Nara, and by loop back configuration; in total 635 km of transmission was performed. As discussed in Chapter 2, the CS-RZ modulation format was immune to the fiber nonlinearity because there was no intense carrier in the spectrum. The transmission loss for one span was 15 dB, and dispersion was 10 ps/nm. To compensate for the loss, EDFA were put for each span. The total dispersion was adjusted to 0 ps/nm with DCF at the input and output ends. The dispersion slope was compensated to almost 100 %. Also used was a PMD (polarization mode dispersion) compensator at the receiver side. The configuration of the PMD compensator is illustrated in Figure 7.5, and it was controlled manually from time to time. At the receiver side, a 40-Gb/s clock signal was extracted using an optoelectronic phase lock loop consisting of an EAM modulator and electronic phase-lock loop circuit. The root mean square time jitter of the extracted clock was about 60 fs. The DEMUX was done by EAM driven by the clock recovery circuit with an RF amplifier. The EAMs used here were designed to respond to arbitrary polarization.

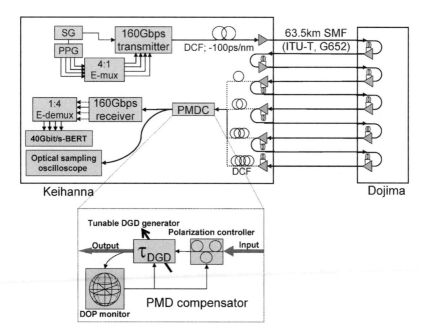

Figure 7.5   Experimental setup for field transmission experiments using JGNII test bed by Murai *et al.* (Reproduced by permission of © 2006 National Institute of Information and Communications Technology (NICT) [3])

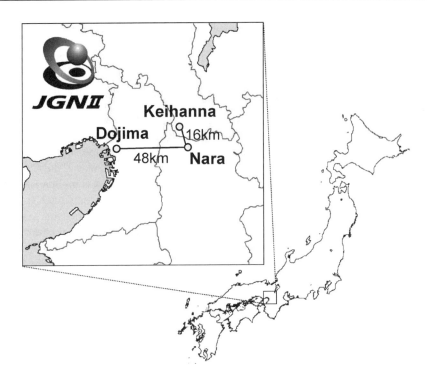

Figure 7.6 Map for the location transmission experiment. Transmission was done between Keihanna and Dojima via Nara. (Reproduced by permission of © 2006 National Institute of Information and Communications Technology (NICT) [3])

Figure 7.7 shows the Q-factor for four 40-Gb/s tributary channels after various transmission distances. Although the Q-factor deteriorates over longer transmission spans, the value was 15.7 dB after 635 km transmission, which corresponds to an error rate of 10^{-9}. Through this experiment, the usefulness of EAM-based OTDM light sources capable of generating CS-RZ signals was demonstrated.

7.2.3 Recent Transmission Experiments above 160-Gb/s

To see the up-to-date trend of high-speed transmissions, we review the experiments reported in ECOC over the past three years (2005–2007). We picked up papers in which devices used in experiments are described in some detail.

Table 7.1 summarizes the reported experiments. Of interest in the table are various modulation formats and multiplication schemes used in experiments. Modulation formats are OOK (on-off keying), CS-RZ (carrier suppressed return to zero), DPSK (differential phase shift keying), DQPSK (differential quadrature phase shift keying). With the use of a CS-RZ format, we can decrease the effect of the fiber nonlinearity effect because of there is no intense carrier in the spectrum. In the DPSK, we can increase the receiver sensitivity by one-bit delayed homodyne detection. In the DQPSK, we can double the data rate, i.e., this is a two-level transmission. Also used was polarization multiplexing. The polarization state of the light can be considered to be one of the most important resources for higher data rate transmission. The highest data rate of

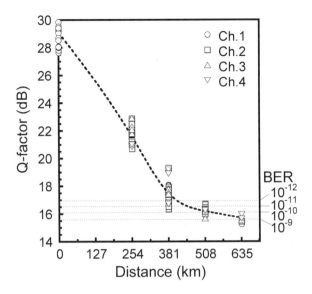

Figure 7.7 Quality factor for transmission over various distances. Even for 635-km transmission, a quality factor of 15.7 dB, which corresponds to an error rate of 10^{-9}, was maintained. (Reproduced by permission of © 2006 National Institute of Information and Communications Technology (NICT) [3])

Table 7.1 Over 160-Gb/s transmission experiments picked up from ECOC 2005–2007. Papers with some description of used optical devices are reviewed

Year	Bit-rate, distance, etc.	Format and scheme	Devices and equipments used in the experiments		Reference
			Transmitter	Receiver	
2005	160 Gb/s 8 WDM, 200-km field	RZ-DPSK, OTDM	Two-stage EAM for pulse generation, LN mod.	DEMUX by two-stage EAM, Balanced detector, PMD-compensator	[4]
2005	170 Gb/s 8 WDM 421-km field	CS-RZ with FEC, OTDM	DFB-LD, MZ-mod. and EAM	DEMUX by EAM, PMD-compensator	[5]
2005	640 Gb/s 480 km	DQPSK, OTDM Pol.-MUX.	LD-pulse source LN-MZ phase mod. LN phase mod.	DEMUX by EAM, DQPSK receiver, Balanced detector	[6]
2005	1.28 Tb/s 240 km 2.56 Tb/s 160 km	DQPSK, OTDM Pol. mux	ML- solid state laser, Pulse compressor, LN- MZ phase mod.	DEMUX by NOLM drived by ML-fiber laser	[7]

(*continued overleaf*)

Table 7.1 (*continued*)

Year	Bit-rate, distance, etc.	Format and scheme	Devices and equipments used in the experiments		Reference
			Transmitter	Receiver	
2006	160 Gb/s 600 km	OOK, OTDM, OFT	ML-Fiber laser, LN mod.	DEMUX by EAM, OFT using LN-mod.	[8]
2007	160 Gb/s 900 km	DPSK OFT OTDM	ML-Fiber laser, LN mod.	DEMUX by EAM, OFT using LN-mod.	[9]
2007	160 Gb/s 100 km	OOK OTDM	ML-Fiber laser	DEMUX by two-stage EAM	[10]
				λ-conversion by SOA with filtering	

Abbreviations: Pol. Mux: polarization multiplexing; ML: mode locked; LN-MZ mod.: LiNbO$_3$-Mach–Zehnder modulator

2.56 Tb/s for a single channel was demonstrated by Weber *et al.* using DQPSK, OTDM and polarization multiplexing [7]. A new very interesting transmission scheme was also reported by Hirooka *et al.* [8, 9]. They used an optical Fourier transform (OFT) method for the transmission. This scheme is based on the fact that even when an optical pulse shape is deformed in the time domain by linear dispersion of fiber, the spectral shape is not affected by transmission. By using this fact, the original pulse shape can be recovered after transmission by use of the time-domain optical Fourier transform technique. Using this scheme, a 160 Gb/s, 900-km transmission was demonstrated. In some experiments, forward error correction (FEC) is employed [5]. This technology reduces the stringent requirement of error rate.

7.3 Requirements on Devices and Prospects

Transmission experiments reviewed so far demonstrated that we can perform very high data rates up to 2.56-Tb/s transmission over several hundred kilometers. This is a great achievement in research and development of ultrafast devices and transmission technologies. However, it is as yet far from a stage of deployment for commercial communications systems. The cost and size of equipment are the problem. This comes mainly from the as yet insufficient device development. In the following we discuss further problems to be solved in the devices together with their prospects, and also discuss the necessity for new technologies.

7.3.1 Devices Described in this Book

7.3.1.1 Mode Locked Lasers (Chapter 2)

Mode-locked laser diodes are sometimes used for short pulse generation of 40-Gb/s. The hybrid mode-locking operation mode can generate short pulses of around 1–2 ps with jitter less than 0.18 ps at a 40-Gb/s range. Also demonstrated was subharmonic synchronous mode locking in colliding pulse mode-locked laser diodes, which realized a 160-Gb/s repetition operation with a small jitter. As an application, clock extraction from a distorted 160-Gb/s signal was demonstrated in Chapter 2. One shortcoming, however, could be control of the repetition rate, which is determined by the cavity length. Some means to realize tunable repetition rate mode-locked LD should be exploited to make this device more attractive.

Although the mode-locked LD is an indispensable device for ultrafast systems, fiber ML lasers were used in many experiments as listed in Table 7.1. One reason is the easy repetition rate control in fiber based ML-lasers. However the major cause for this would be that it was difficult to purchase the devices. Only a limited number of companies can supply this device. Increasing the research into ultrafast technologies will increase the users of the device and then the situation will be improved.

7.3.1.2 EAM and EAM-based Light Source and DEMUX (Chapter 2)

EAM was used to generate short pulses as demonstrated in the generation of CS-RZ signals in Chapter 2, and was applied to field transmission experiments. EAM was also used for DEMUX operation in many transmission experiments as can be seen from Table 7.1. A shortcoming is the insertion loss, which is typically a few 10 dB. It was necessary to introduce EDFA to compensate for the loss. For processing very short pulses, below 3 ps, a two-stage cascaded configuration is needed. Here again, additional amplification is required and an additional RF-drive circuit. Although the EAM is a very convenient device for transmission experiments, the drawbacks are insertion loss and the necessity for microwave circuits for operation. As for the insertion loss, integration with SOA by hybrid integration technology either using PCL or an Si-wire waveguide could be a promising solution for this shortcoming.

7.3.1.3 SOA Based Devices (Chapter 3)

The SOA is most frequently used for ultrafast signal processing. In the FESTA experiments, a hybrid integrated SMZ gate based on the optical nonlinearity of SOA was used for DEMUX operation. As described in Chapter 2, the SMZ gate demonstrated versatile application for signal processing, DEMUX operation, wavelength conversion, 2R operation, and also the operation for NRZ signals. The DEMUX operation of a 640-Gb/s signal to 40-Gb/s and 10-Gb/s was demonstrated. A shortcoming, however, could be that there are many control parameters for the operation, i.e., currents to two SOAs, precise phase shift adjustment by heater current to the arm of the Mach–Zehnder interferometer. Improvements for this would make this device more attractive.

The use of a wavelength filter to extract only the ultrafast response component of the SOA response as discussed in Chapter 3 is one very simple and attractive way of using SOA. As referred in Chapter 3 and listed in Table 7.1, many applications have been reported. SOA will continue to be an essential device for ultrafast signal processing.

7.3.1.4 UTC-PD/TW-EAM Devices (Chapter 4)

For this device, DEMUX operation of 320 Gb/s to 10 Gb/s, 100-Gb/s wavelength conversion, and 100-Gb/s error-free retiming operations were demonstrated. A unique feature is that this device does not use optical nonlinearity directly as do other devices. The key to higher speed operation is the reduction of the RC limit and phase matching at the TW-EAM. In this device, we can get rid of the intrinsic trade-off relation of optical power and response speed as discussed in Chapter 1. A unique feature of this device is that it can cover a bit rate of around 100 Gb/s, where devices using the ultrafast nonlinear response, for example the ISBT gate, are not suited because of their too fast response. The bit rate of 100 Gb/s is of importance as the next standard of the bit rate of ethernet. As discussed in Chapter 4, this device can also be used at the still higher bit-rate of more than 320 Gb/s by improvement of the RC and the phase matching.

7.3.1.5 ISBT Gate (Chapter 5)

This device is still under development and has not yet been used in transmission experiments. There are many issues to be overcome for practical application. The first is the high optical energy needed to saturate the absorption. This is directly related to the ultrafast response (200 fs–1 ps). There is, however, still room for a lower operating energy by improving the devices, crystal quality, quantum well structure and waveguide structure. Secondly, there is essentially a large insertion loss in the absorption–saturation type operation. This drawback is the same as in the case of EAM. Thirdly, there is unavoidable polarization dependence, because the intersub-band transition takes place only for TM polarization. Polarization dependence, however, could be overcome by polarization diversity configuration. Despite these unresolved problems and drawbacks, this device is highly attractive for future ultrafast systems because of the intrinsic ultrafast response of the intersub-band transition.

The newly found operation mode for the InGaAs/AlAs/AlAsSb ISBT gate, i.e., phase modulation takes place for loss-less TE mode by TM control pulse, is highly promising for practical devices. This allows us to realize low-insertion loss ultrafast devices. This could be one of the breakthroughs for the realization of practical ISBT gate devices.

7.3.1.6 Wavelength Converter (Chapter 6)

FWM wavelength converters based on SOAs and LDs are described in this book. Wavelength conversions for short pulses and 160-Gb/s signals were demonstrated. The advantage of SOA-based FWM wavelength conversion is in its large bandwidth and resultant transparent nature. Using a two-wavelength, pump scheme, wavelength conversion to replicate was demonstrated for both 160-Gb/s OOK and DPSK. This demonstrates the transparent nature of the wavelength conversion. A shortcoming is the problem of asymmetric conversion efficiency with respect to the pump wavelength. As demonstrated, this can be overcome by the use of a quantum dot SOA or by use of a two-wavelength pumping scheme. Integration with tunable pumping light sources will result in highly attractive compact devices.

7.3.2 Necessity for New Functionality Devices and Technology

7.3.2.1 Phase Modulator

In recent transmission experiments, phase modulation such as PSK, DQPSK have been used. The multivalue scheme using the phase of light is becoming an important and promising scheme. For this purpose, an efficient phase modulator is an essential device. Also, the optical

Fourier transform scheme requires aphase modulator [8, 9]. So far, LN-based phase modulators have been used as listed in Table 7.1. We have not yet realized a good semiconductor-based phase modulator. The realization of large refractive index modulation has long been a dream for the optical semiconductor device researcher. If we could do this, we could create not only a phase modulator but also a large tuning range of tunable lasers with a simple configuration. One possibility for this, though restricted in the very high bit-rate region, is the InGaAs/AlAs/AlAsSb ISBT gate device as described in Chapter 5. This device provides a pure, all-optical phase modulation at 1 ps response speed for TE probe light by TM control pulse. However, its amount of the phase shift does not reach π. This device is worth further research and development toward the realization of large phase modulation.

7.3.2.2 Polarization Compensation Devices

In field transmission experiments, PMD compensation is very important. For example, in the transmission experiment using a JGNII test bed, the PMD compensator shown in Figure 7.5 was used [3]. This was controlled manually because the PMD change was due to rather slow environmental change. In real systems there could be a rapid change in PMD due to some shock or mechanical vibration. More advanced PMD compensators should be developed. It is unclear as to whether or not semiconductor-based devices can contribute to developing good PMD compensators.

7.3.2.3 Hybrid Integration Technology

What is important in order to make ultrafast communication a reality is the size of the equipment. At present, a 40-Gb/s transceiver is a size of a few tens of centimeters square. If we plan to commercialize a 160-Gb/s transceiver, its size should never exceed four-times of a 40-Gb/s transceiver, or should be the same as that of a 40-Gb/s transceiver. Transmission equipment so far used in experiments is far from this size. To realize small size equipment, the key technology should be hybrid integration technology. The integration of ultrafast semiconductor devices with waveguide components made of Si-wire or PLC would be a key in this. The hybrid integrated SMZ gate switch, where hybrid integration was done using PLC, was described in Chapter 3. By using the Si-wire waveguide we can realize a much smaller Mach–Zehnder switching gate. The hybrid integration of EAMs and SOAs may miniaturize the CS-RZ light source described in Chapter 2. In hybrid integration, there will be many difficult problems to be overcome, such as coupling of the waveguide and semiconductor device, reflection at the waveguide joint, and long-term stability of the precise alignment, and heat dissipation. Although, the hurdles are high, it is essential to establish the hybrid integration technology for practical ultrafast equipment.

7.4 Summary

Semiconductor-based, all-optical ultrafast signal processing devices are described in this book. Some of the devices have been tested in transmission experiments and their usefulness has been demonstrated. However, from the viewpoint of realizing commercial ultrafast communications networks, many devices presented in this book remain at a stage of challenge. As pointed out in the previous section, there are many issues further to be overcome in the present devices. It is even probable that some of the devices described in this book may not be used in future communications systems. However, the challenges so far have given us precious knowledge and technologies that speed up research and development. We can say that the ultrafast communications systems are not so far off. They will bring benefits of huge-capacity, real-time, transmission in our daily life and in our economy, as discussed in Chapter 1. An important point is that ultrafast technology has the potential to realize a system with low power consumption. The authors of this book believe that some of the devices in this book, or new devices developed hereafter based on the study in this book should play key roles in ultrafast communication systems in the near future.

References

[1] A. Suzuki, X. Wang, T. Hasegawa, Y. Ogawa, S. Arahira, K. Tajima, and S. Nakamura, '8 × 160 Gb/s (1.28 Tb/s) DWM/OTDM unrepeated transmission over 140-km standard fiber by semiconductor-based devices,' *ECOC-IOOC 2003*, Proceedings Volulme 1, pp. 44–47, Rimini, Italy (2003).

[2] A. Suzuki, X. Wang, Y. Ogawa, and S. Nakamura, '10 × 320 Gb/s (3.2 Tb/s) DWDM/OTDM transmission in C-band by semiconductor-based devices,' *ECOC 2004 Proceedings,* Post-Deadline Paper, Th4.1.7, pp. 14–15, Stockholm, Sweden (2004).

[3] H. Murai, 'EA modulator based OTDM technique for 160 Gb/s optical signal transmission,' *Journal of the National Institute of Information and Communication Technology*, **53**(2), 27–35 (2006).

[4] M. Daikoku, T. Miyazaki, I. Morita, H. Tanaka, F. Kubota, and M. Suzuki, '160 Gb/s-base field transmission experiments with single-polarization RZ-DPSK signals and simple PMD compensator,' *ECOC 2005 Proceedings*, Volume 3, pp.375–378, paper We2.2.1, Glasgow (2005).

[5] S. Vorbeck, M. Schmidt, R. Leppla, W. Weiershausen, M. Schneiders, and E. Lach, 'Long haul field transmission experiment of 8 × 170 Gbit/s over 421 km installed legacy SSMF infrastructures,' *ECOC 2005 Proceedings*, Volume 3, pp. 432–435, paper We3.2.1, Glasgow (2005).

[6] S. Ferder, C. Schubert, R. Ludwig, C. Boemer, C. Schmidt-Langhorst, H. G. Weber, '640 Gb/s DQPSK single-channel transmission over 480 km fibre link,' *ECOC 2005 Proceedings*, Volume 3, pp. 437–438, paper We3.2.2, Glasgow (2005).

[7] H. G. Weber, S. Ferder, M. Kroh, C. Schmidt-Langhorst, R. Ludwig, V. Marembert, G. Boemer, F. Futami, S. Watanabe, and C. Schubert, 'Single channel 1.28 Tb/s and 2.56 Tbit/s DQPSK transmission,' *ECOC 2005 Proceedings,* Volume 6, pp. 3–4, paper Th4.1.2, Glasgow (2005).

[8] T. Hirooka, K. Hagiuda, T. Kumakura, K. Osawa, and M. Nakazawa, '160 Gb/s-600 km OTDM transmission using time-domain optical Fourier transformation,' *ECOC2006 Proceedings*, Volume 2, pp. 31–32, paper Tu 1.5.4, Cannes (2006).

[9] T. Hirooka, M. Okazaki, K. Osawa, and M. Nakazawa, '160 Gbit/s-900 km DPSK transmission with time-domain optical Fourier transformation,' *ECOC 2007 Proceedings,* Volume 1, pp. 55–56, Berlin (2007).

[10] J. Herrera, O. Raz, Y. Liu, E. Tangdiongga, F. Ramos, J. Marti, H. de Waardt, A. M. J. Koonen, G. D. Khoe, H. J. S. Dorren, '160 Gb/s error-free transmission through a 100-km fibre link with mid-span all-optical SOA-based wavelength conversion,' *ECOC2007 Proceedings*, Volume 1, pp. 57–58, Berlin (2007).

Index